Engaging *with* Nature

Engaging *with* Nature

Essays on the Natural World in Medieval and Early Modern Europe

Barbara A. Hanawalt

and

Lisa J. Kiser

editors

University of Notre Dame Press
Notre Dame, Indiana

Copyright © 2008 by University of Notre Dame
Notre Dame, Indiana 46556
www.undpress.nd.edu
All Rights Reserved

Published in the United States of America

Library of Congress Cataloging-in-Publication Data

Library of Congress Cataloging-in-Publication Data

Engaging with nature : essays on the natural world in medieval and early modern Europe / editors, Barbara A. Hanawalt and Lisa J. Kiser.
p. cm.
Includes bibliographical references and index.
ISBN-13: 978-0-268-03083-4 (pbk. : alk. paper)
ISBN-10: 0-268-03083-9 (pbk. : alk. paper)
1. Philosophy of nature–Europe–History. I. Hanawalt, Barbara. II. Kiser, Lisa J., 1949–
BD581.E54 2008
304.209—dc22

2008009049

∞ *The paper in this book meets the guidelines for permanence and durability of the Committee on Production Guidelines for Book Longevity of the Council on Library Resources.*

Contents

Acknowledgments   vii

Introduction   1
Barbara A. Hanawalt and Lisa J. Kiser

CHAPTER ONE
Homo et Natura, Homo in Natura:
Ecological Perspectives on the European Middle Ages   11
Richard C. Hoffmann

CHAPTER TWO
Inventing with Animals in the Middle Ages   39
Jeffrey Jerome Cohen

CHAPTER THREE
Ritual Aspects of the Hunt *à Force*   63
Susan Crane

CHAPTER FOUR
The (Re)Balance of Nature, ca. 1250–1350   85
Joel Kaye

CHAPTER FIVE
Collecting Nature and Art: Artisans and Knowledge
in the *Kunstkammer*   **115**
Pamela H. Smith

CHAPTER SIX
"Procreate Like Trees": Generation and Society in
Thomas Browne's *Religio Medici*   **137**
Marjorie Swann

CHAPTER SEVEN
Human Nature: Observing Dutch Brazil   **155**
Julie Berger Hochstrasser

Bibliography   **201**

Contributors   **225**

Index   **227**

Acknowledgments

We express our gratitude to Ohio State University's Center for Medieval and Renaissance Studies and to its College of Humanities for supporting the series of lectures and colloquia in 2004–2005 that ultimately resulted in this volume. In connection with that series, we thank Pat Swinehart for helping to make it run smoothly. We are also grateful to Barbara Hanrahan of the University of Notre Dame Press and to the two thoughtful readers she assigned to review the volume in its earlier drafts. Thanks, too, to James L. Battersby for his tireless improvement of our volume's ideas and the prose that expresses them. Finally, our gratitude also extends to Ryan Judkins for help with the index and the editorial assistants at the University of Notre Dame Press, especially Rebecca DeBoer, for their editorial wisdom and good cheer at various stages of the production process.

Introduction

Barbara A. Hanawalt and Lisa J. Kiser

The seven essays in this collection address the subject of the natural world in a number of medieval and early modern contexts, each giving special attention to human interactions with the natural environments that surrounded and supported both life and culture. These essays, representing several disciplines, and sometimes combining traditional disciplines, are designed to make readers aware of current scholarship investigating nature in the premodern and early modern past.

Special problems beset historians and cultural critics treating the nonhuman natural world in these periods. The most daunting problem is that in both the visual and written records of the time, nature appears to be both everywhere and nowhere. In the broadest sense, of course, nature is everywhere. It supplies the most important contexts for human survival, since agriculture, animal husbandry, medicine, and the patterns of human settlement and migration all have their basis in natural settings. Moreover, humans marked personal, community, daily, and seasonal events by natural occurrences and built their cultural explanations around the workings of nature, which formed the unspoken backdrop for every historical event and document of the time. Nature was everywhere, too, in the texts and artifacts in which medieval and early modern people recorded and represented their social and religious identities. The surviving records relating to heraldry, hunting, cooking, theology, folklore, sports, science, and art (to mention only a few of the cultural arenas in which nature features prominently) vividly reflect medieval and early modern responses to the natural world. Despite the ubiquity of nature's presence in the artistic and literary cultures of these periods, however, it is also true that overt discussion of it is seldom found. Until the sixteenth century, when scientific writing began to take as its primary subject the close observation of nature, responses to nature were often recorded only in the course of an investigation of some other subject, such as how one might interpret a certain biblical passage, or concoct an effective treatment for fever, or ensnare a rabbit, or (to take an Aesopic example) satirize irresponsible clergy. In other words, medieval and early modern writers, unlike their modern counterparts, seldom sat down to write essays or treatises on nature in and of itself. Indeed, for many writing in these early periods, "nature" was arguably not even a discursive category; it simply went without saying.

Consequently, modern scholars seeking to analyze the understanding of nature in the medieval and early modern periods often find it necessary to become experts in fields seemingly unrelated to their central concern. They may need to become familiar with theology; social history; literary and other artistic forms; agricultural, medicinal, and culinary traditions; or any number of practices that necessarily involve nature or attitudes about nature: war, sports, hunting, divination, pet-keeping, law, and private devotion, to name a few. Thus, medieval or early modern natural and environmental history, perhaps more than most modern academic specializations, is profoundly interdisciplinary, requiring attention not only to the aspect of nature under analysis, but also to the social, philosophical, and scientific context in which it is found. This observation finds ample illustration in the essays in this volume. Each skillfully weaves together knowledge from disparate fields to gain insight into an aspect of nature as it was understood or experienced in medieval and early modern Europe. The special topics covered here include animal/human relationships, environmental and ecological history, medieval hunting, early modern collections of natural objects, the moral relationship of religion and nature, the rise of science, and the motives underlying the artistic representations of plants, animals, and humans made by Europeans encountering the New

World for the first time. Although wide-ranging in their approaches, the essays in this volume also contribute to traditional disciplines, such as the history of art, the history of science, environmental history, literary history, political history, and the history of ideas. The study of nature in the medieval and early modern periods is an emerging discipline in its own right, but the essayists also add to bodies of knowledge defined by the distinctive standards, methodologies, and paradigms of established fields.

To situate the essays in current research, it is useful to mention briefly two matters critical to understanding the issues at stake. The first is the problematic ontological status of the category of "nature" itself; the second is the place that this collection assumes in the larger history of scholarship on medieval and early modern thinking about the nonhuman natural world. Neither of these important topics can be adequately addressed in a short introduction: this is not the place to conduct lengthy philosophical arguments or to survey the large (and growing) body of academic work that is the context for our project. Yet both of these matters—the various philosophical assumptions informing the following essays and the history of scholarship that has made possible our essayists' analyses—are the crucial "background conditions" on which this collection depends. It is thus appropriate that we briefly address them here.

With respect to the philosophical issue, we realize that any attempt to talk about "nature" in an interdisciplinary context will always be a vexed one, since various disciplines, and even various practitioners within them, differ substantially in the ways they conceptualize "nature" and deploy the word to signify their conceptualizations. For those identifying themselves with poststructuralist critique, for example, first among their concerns is the status of what we casually term "nature" itself. Is "nature" just as socially constructed as "culture"? As the anthropologist Marilyn Strathern has framed the question, has the one term (culture) consumed the other term (nature), so that the traditional antithesis between nature and culture has become occluded? Or, is it rather the case that in the course of investigating how activities and attitudes in relation to nature alter or remain the same through time, "nature" can usefully be posited as "an enduring, even timeless, phenomenon"?[1] The essayists in this volume approach their material from a number of positions with respect to what is, for theorists, a troubled and porous nature/culture divide. For some of our authors, nature is "fact," while for others, nature is clearly "value"; still others, it appears, would locate nature somewhere along a spectrum between the two. But some might also wish to reject both the binary and the spectrum altogether. Because of such conceptual and methodological diversity in the work of the contributors, this volume is stronger than if it had been limited to a single theoretical perspective.

Next, in considering this volume's place in the context of scholarship on nature in the medieval and early modern periods, we see ourselves and our authors as heirs to a large body of knowledge produced by distinguished scholars, beginning with historians of science who studied the growth of natural philosophy in the medieval and early modern eras. Edward Grant's and David Lindberg's books on the simultaneous development of religion and science in postclassical Europe are foundational, for example, but so are R. G. Collingwood's and, later, George Economou's more humanistic reflections on the history of the concept of nature in European thought.[2] Keith Thomas's work on early modern attitudes toward the natural world still stands as an important witness to what can be inferred from the historical record.[3] The study of classical, medieval, and early modern natural histories, from Pliny to Gesner, has resulted in scholarly inquiry on the roles played by specific plants, animals, minerals, and other natural objects in the cultural and physical lives of medieval and early modern people, as have modern analyses of bestiaries, lapidaries, and herbals.[4] A volume such as this is also indebted to historical ecologists, who have studied climate patterns, pollen residues, buried bones, ancient trees, scars on the landscape, and other physical evidence from the earth's historical past.[5]

Apart from the specialized studies cited in the essays in this volume, numerous recent works have contributed to the general study of nature in medieval and early modern Europe. Alfred W. Crosby's *Ecological Imperialism: The Biological Expansion of Europe, 900–1900* (1986); Francis Klingender's *Animals in Art and Thought to the End of the Middle Ages* (1971); Robert Delort's *Les animaux ont une histoire* (1984); Carolyn Merchant's *The Death of Nature: Women, Ecology, and the Scientific Revolution* (1980); Clarence J. Glacken's *Traces on the Rhodian Shore: Nature and Culture in Western Thought from Ancient Times to the End of the Eighteenth Century* (1967); and Joyce E. Salisbury's *The Beast Within: Animals in the Middle Ages* (1994) have all been instrumental in shaping the study of nature in modern scholarship. For postmodern theorizing on the animal/human relationship in general, readers may consult the work of Richard Sorabji, Cary Wolfe, and Nigel Rothfels; and for the study of landscapes in their relationship to culture and politics, the informative collection edited by W.J.T. Mitchell.[6] For those who wish to further their knowledge about the specific topics covered in this volume, we have included a bibliography of the works cited in the essays, which also provides readers with a good sense of the new directions in which the study of nature in the medieval and early modern periods is heading. The essays in this collection, we believe, carry forward and vividly exemplify some of these new directions.

Richard C. Hoffmann's study of the interrelationship of medieval people and their natural environment aptly serves as the volume's opening essay, for it reminds us that the possibilities and limitations of human history in general are largely shaped by natural forces such as climate patterns and their accompanying cycles of drought, famine, and disease. Yet he shows that human populations were not merely passive victims of natural forces but also active collaborators with nature in the unfolding of history. Humans have manipulated and modified the natural world with, for example, their complex, socially ordered food chains and their practices of hunting, animal husbandry, and woodland clearance. Combining environmental and social history, Hoffmann's essay forcefully argues that humans (with their interests, their beliefs, and their desires) have worked together with nonhuman natural forces to effect change in both the human and the natural spheres.

Using an innovative cultural studies approach, Jeffrey Jerome Cohen shows how medieval people often created "thought experiments" in which animals were used as vehicles to imagine worlds different from their own—or worlds and behaviors prohibited by their own. Bestiary lore involving animal reproduction could often serve, for example, as a way to "safely" discuss human sexual practices. Similarly, discussion of human race and ethnicity (blackness or Jewishness, for example) was often carried on by means of animal stories that pondered the results of "cross-breeding" or hybridity. In short, as Cohen argues, representations of animals in human bodies and humans in animal bodies allowed for subtle medieval theorizing about identity and social relations.

Continuing the theme of animal/human relations, Susan Crane's essay demonstrates exactly how and why the aristocracy of medieval Europe made hunting a ritualized display of their social superiority. Crane argues that medieval hunting practices were developed to emphasize the social dominance of their users and outlines the ways in which the animals involved in the hunt (horse, hound, stag, boar) were "nobilized" and then choreographed in the hunt's elaborate staging. The essay also discusses the special partnership formed by humans and their hunting dogs, which included a complex "language" to ensure cross-species communication. The close partnership between human and animal, with the human in control, signaled to others the social dominance of the aristocrats over nature (as well as, implicitly, over culture).

Joel Kaye, in a compelling new argument in the history of science, discusses the ways in which thirteenth- and fourteenth-century medieval scholastics (Jean Buridan, in particular) began to think of nature as a system displaying equilibrium in the relationship of its parts to the whole. Buridan's geological writings, which exemplify this novel, non-Aristotelian and nonscriptural view of the natural world, suggest that the earth undergoes physical changes through time, with each change resulting in an adjustment elsewhere in the system. Buridan and his colleagues were surely unaware of the radical novelty of their analyses. But by taking a retrospective view, Kaye demonstrates how their work on geology constituted a major contribution to the development of science.

Pamela H. Smith's essay, on the topic of early modern collections of natural and man-made objects, engages fields as diverse as art history, political theory, and the history of science. Smith argues that sixteenth-century private collections of natural objects have important connections to social history and to the history of the philosophy of nature. European collections from this period include natural objects, such as animal parts (beaks, claws, feathers, and so forth) and other specimens (seeds, fluids from trees, dyes, and clays), but also "realistic" man-made statues of animals and samples of human prostheses (such as false limbs) shaped from animal parts. In analyzing these intriguing juxtapositions of natural and man-made objects, Smith demonstrates that sixteenth-century collectors valued the imitation of natural processes and objects and, furthermore, that artisanal knowledge, such as that manifested in the incredible skill with which craftsmen could replicate "life," was an important aspect of early modern attempts to assert human control over the natural world.

Focusing on seventeenth-century British thought, Marjorie Swann argues that biological ideas were crucial to the work of Sir Thomas Browne, an important early modern religious moralist and social theorist. One of his enduring ideas was that human sexual relations are unseemly and morally imperfect, because they depart from God's creation of Adam and Eve in Eden, a creation involving no sexuality at all. Browne's natural models for this asexual form of reproduction are trees, which rely merely on proximity—and the gentle breezes—for their reproductive success. Swann suggests that Browne's admiration for this biological model influenced him to venerate friendship, rather than heterosexual marriage and procreation, as the best possible communion between humans. Her essay makes a major statement about an important figure in British intellectual history, and it does so by bridging the disciplines of literary analysis and the history of biological science.

Finally, Julie Berger Hochstrasser's essay persuasively argues that artistic representations of the natural world are deeply dependent on who is looking, what is being looked at, and why the observation is taking place. In a richly detailed interpretation of the ways Europeans created and viewed images of wildlife and plants from the new Portuguese colony of Brazil, Hochstrasser focuses specifically on pictures of the sloth, an animal alien to Europe and one frequently represented by visitors to the New World. Which of the extant representations come from "life"? Which are copied from books? Are the animals and the peoples of Brazil represented as parts of larger ecosystems, habitats, and social systems, or are they willfully extracted from these and made independent "aesthetic" objects of contemplation? Her essay raises (and answers) major questions concerning the colonial occupation of Brazil, in an intriguing synthesis of anthropology, biology, art history, and the history of European colonialism.

Current historians of nature—whether approaching their topic through interdisciplinary cultural studies, social and economic history, the history of science and philosophy, or traditional literary analysis—rightly regard their subject as a new one. Nonetheless, a case can be made that the subject first achieved legitimacy in the early 1960s, when Rachel Carson's work showed the immediate practical value of studying and analyzing humanity's interactions with nature. Following the appearance of *Silent Spring* (1962), scholarship on environmental thought took on a new urgency, often aligning itself with concerns about the ways in which human social and economic practices, both past and present, have negatively affected the natural processes that sustain life on earth. The essayists in this volume do not directly address the current political climate. Each, however, contributes to the project of documenting the significant human responses to the earth and its denizens that have shaped and still shape many of our beliefs about, and attitudes toward, nature. Historical views of nature continue to affect both individual action and public policy with respect to the environment. Environmental history, then, in the words of two of its current practitioners, "has the promise to be central to the most influential social thought in the academy and among policy makers."[7] This volume provides material for reflection on our own relationships to nature at a time when scrutiny of those relationships is crucial to our continued existence.

Notes

1. Marilyn Strathern, *After Nature: English Kinship in the Late Twentieth Century* (Cambridge: Cambridge University Press, 1992), 2 and 5. Strathern's conceptualization of the "nature/culture" problem is complex and deserves careful attention. See also Lorraine Daston and Fernando Vidal's "Introduction: Doing What Comes Naturally," in *The Moral Authority of Nature* (Chicago: University of Chicago Press, 2004), 1–20. For an argument addressing the conceptual poverty of the "nature/culture" distinction, see Bruno Latour, *The Politics of Nature: How to Bring the Sciences into Democracy* (Cambridge, Mass.: Harvard University Press, 2004).

2. Edward Grant, *The Foundations of Modern Science in the Middle Ages: Their Religious, Institutional, and Intellectual Contexts* (Cambridge: Cambridge University Press, 1996), and *God and Reason in the Middle Ages* (Cambridge: Cambridge University Press, 2001); David C. Lindberg, *The Beginnings of Western Science: The European Scientific Tradition in Philosophical, Religious, and Institutional Context, 600 B.C. to A.D. 1450* (Chicago: University of Chicago Press, 1992); R. G. Collingwood, *The Idea of Nature* (Oxford: Clarendon Press, 1945); and George D. Economou, *The Goddess Natura in Medieval Literature* (1972; repr., Notre Dame, Ind.: University of Notre Dame Press, 2002). Also useful are the two collections edited by Arjo Vanderjagt and Klaas van Berkel, *The Book of Nature in Antiquity*

and the Middle Ages (Leuven: Peeters, 2005) and *The Book of Nature in Early Modern and Modern History* (Leuven: Peeters, 2006).

3. Keith Thomas, *Man and the Natural World: Changing Attitudes in England, 1500–1800* (Harmondsworth: Penguin, 1983).

4. For a useful account of the study of natural history from antiquity to the early modern period, see Brian W. Ogilvie, *The Science of Describing: Natural History in Renaissance Europe* (Chicago: University of Chicago Press, 2006). For the medieval bestiaries, see Florence McCulloch, *Medieval Latin and French Bestiaries* (Chapel Hill: University of North Carolina Press, 1960); Wilma George and Brunsdon Yapp, *The Naming of the Beasts: Natural History in the Medieval Bestiary* (London: Duckworth, 1991); Debra Hassig, *Medieval Bestiaries: Text, Image, Ideology* (Cambridge: Cambridge University Press, 1995); Ron Baxter, *Bestiaries and Their Users in the Middle Ages* (Phoenix Mill: Sutton, 1998); Willene B. Clark, *A Medieval Book of Beasts: The Second Family Bestiary* (Woodbridge: Boydell Press, 2006); and Jacques Voisenet, *Bêtes et hommes dans le monde medieval: Le bestiare des clercs du Ve au XIIe siècle* (Turnhout: Brepols, 2000). On medieval and early modern botanical study, see Frank J. Anderson, *An Illustrated History of the Herbals* (New York: Columbia University Press, 1977), and Minta Collins, *Medieval Herbals: The Illustrative Tradition* (Toronto: University of Toronto Press, 2000).

5. Good accessible examples of historical ecology are the studies of English woodlands by Oliver Rackham; see especially his *The History of the Countryside* (London: Dent, 1986) and *Trees and Woodlands in the British Landscape: The Complete History of Britain's Trees, Woods & Hedgerows* (London: Dent, 1976; rev. ed. 1990). For an understanding of the variety of work that historical ecologists do, see Emily W. B. Russell, *People and the Land Through Time: Linking Ecology and History* (New Haven: Yale University Press, 1997); Carole L. Crumley, ed., *Historical Ecology: Cultural Knowledge and Changing Landscapes* (Santa Fe: School of American Research Press, 1994); David Arnold, *The Problem of Nature: Environment, Culture and European Expansion* (London: Blackwell, 1996); Robert Delort and Francois Walter, *Histoire de l'environnement européen* (Paris: Universitaires de France, 2001); and John McNeill, "Observations on the Nature and Culture of Environmental History," *History and Theory* 42 (2003): 5–43.

6. Richard Sorabji, *Animal Minds and Human Morals: The Origin of the Western Debate* (London: Duckworth, 1993); Cary Wolfe, ed., *Zoontologies: The Question of the Animal* (Minneapolis: University of Minnesota Press, 2003); Cary Wolfe, *Animal Rites: American Culture, the Discourse of Species, and Posthumanist Theory* (Chicago: University of Chicago Press, 2003); Nigel Rothfels, ed., *Representing Animals* (Bloomington: Indiana University Press, 2002); W. J. T. Mitchell, ed., *Landscape and Power*, 2nd ed. (Chicago: University of Chicago Press, 2002). For a more traditional view of the cultural and artistic functions of literary and pictorial landscape, see Derek Pearsall and Elizabeth Salter, *Landscapes and Seasons of the Medieval World* (Toronto: University of Toronto Press,

1973). Recent important titles on animals in the English medieval and early modern periods include Dorothy Yamamoto's *The Boundaries of the Human in Medieval English Literature* (Oxford: Oxford University Press, 2000); Erica Fudge's *Perceiving Animals: Humans and Beasts in Early Modern English Culture* (Basingstoke: Palgrave, 2000) and *Brutal Reasoning: Animals, Rationality, and Humanity in Early Modern England* (Ithaca, N.Y.: Cornell University Press, 2006); and Bruce Boehrer's *Shakespeare Among the Animals: Nature and Society in the Drama of Early Modern England* (Basingstoke: Palgrave, 2002). For medieval and early modern environmental history, for example, see Tom Williamson, *Shaping Medieval Landscapes: Settlement, Society, Environment* (Macclesfield, Cheshire: Windgather Press, 2003), and John F. Richards, *The Unending Frontier: An Environmental History of the Early Modern World* (Berkeley: University of California Press, 2003).

    7. Sverker Sörlin and Paul Warde, "The Problem of Environmental History: A Re-Reading of the Field," *Environmental History* 12 (2007): 107.

CHAPTER ONE

Homo et Natura, Homo in Natura

Ecological Perspectives on the European Middle Ages

Richard C. Hoffmann

On the relationship between nature and humankind, most literate medieval Europeans shared certain basic ideological assumptions. *Homo*, "Mankind," was separate and distinct from *Natura*, "Nature." *Homo* in fact had been created to *rule over* Nature, the earth, Creation. Both *Natura* and *Homo* exist temporally in the world of change, the sublunar sphere in a Ptolemaic construction of an Aristotelian universe. Yet from that consensus different medieval thinkers drew implications as divergent as today's stereotypical *Homo ecologicus*, conserving steward, and *Homo devastans*, planetary bane.

The twelfth-century French clerical Neoplatonist Bernard Silvester elaborated in his *Cosmographia* a mythico-scientific allegory of creation and the interrelationship of man (microcosm) and the universe (macrocosm).[1] In the part of his work Bernard devoted to humankind,[2] *Homo* appears as the microcosm, the image of the greater world, who is given knowledge of the ways of things of the world and made "ruler and high priest of Creation, that he may subordinate all to himself, rule on earth and govern the universe."[3] Working with Nature's *tabula fati*, which manage the destiny of all "temporal things subject to change,"[4]

Homo improves the pliant earth provided for him by *Natura*. As the allegorical creatrix Noys put it,

> It is my will that the elements be his, that fire grow hot for him, the sun shine, the earth be fruitful, the sea ebb and flow; that the earth give nourishment to its fruits, the sea to its fish, the mountains to their flocks, and the wilderness to its beasts for him.[5]

Autonomous Nature and humankind thus collaborate to complete and perfect Creation.[6]

A contrasting view was voiced in the early 1490s by Paul Schneevogel (1460/65 to after 1514), a humanist schoolteacher and later municipal official in Upper Saxony who wrote as "Paulus Niavis." His *Iudicium Iovis*, "The Judgement of Jupiter," offered as a parable the vision of a Bohemian hermit in the central European Erzgebirge ("Ore Mountains").[7] The recluse sees the Earth and the classical divinities of Nature together hailing *Homo* before the court of Jupiter, chief of the gods. By mining and woodcutting, *Homo* had injured complaining Earth, who reluctantly yields up her wealth to his coercion. Bacchus, Ceres, the naiads, Minerva, Pluto, Charon, and the fauns report loss and pollution of waters, damage to wine and cereal harvests, destruction of woodlands, noise, and noxious fumes. In his defence, *Homo* replies that Earth was a false mother, withholding love and concealing from him what the gods had provided for the use of humankind.[8] Man's advocates, the dwarves (*penates, wichtellein* in contemporary German-Latin glossaries), assert that *Homo* had to take the entire known world under his protection (*tutela*), which could be done only through work (*labor*) above and below ground. Faced with two strong cases, Jupiter defers judgement to Fortuna, Queen of all things mortal, who declares:

> Men are destined to stab through the mountains, to construct mine shafts, to till fields, to conduct trade, and to strike against the earth, to reject learning; to disturb Pluto; and even to seek out veins of metal in water courses; [man's] body [is destined] to be swallowed up by the earth, to be choked by fumes, intoxicated by wine; to be subjected to hunger and many additional

dangers that it would be best not to know, which are peculiar to humankind.⁹

Humans and Nature thus engage in a costly contest of mutual attrition.¹⁰

Whether *Homo* and *Natura* were in collaboration or conflict, however, both Bernard Silvester and Paul Schneevogel recognized them as together driving *change*, that is, they acknowledged both as part of history, specifically their own history now called "medieval." Their acceptance of natural and human changes is interestingly mirrored by refinements in present-day ecology, the science which studies interactions among organisms and inanimate nature, their mutual agencies, and ways humans can perceive and understand those phenomena. Recent developments of "the new ecology," "historical ecology," and "social ecology" have enlarged this approach by recognizing change and the role human cultures can play in it.¹¹ Natural scientists no longer think of "Nature" as a mere fixed stage set, or even as necessarily tending toward some kind of equilibrium, but rather as encompassing the change essential to history.

Given the awareness shown by medieval writers like Bernard Silvester and Paul Schneevogel, it is perhaps ironic that medieval studies as a field has traditionally been reluctant to admit a nonhuman dynamic to its story. This essay seeks to encourage medievalists to think about the interactions between medieval society and its natural environment and to explore the ecological connections which shaped those changes.¹² Rather than viewing Nature as the passive recipient of human actions, we should acknowledge Nature as an active participant in history, understood as a process of *co-adaptation* of human societies with cultures in their changing environments. Nature changes human society. Human society changes Nature. Evidence from traditional human sources, both verbal and material, and from the growing mass of palaeoscientific data demonstrates this reciprocity in medieval Europe. Those who study medieval life need to possess and contribute to this understanding. Having already illustrated some contemporary grasp of the issue, this essay looks first at a powerful natural dynamic, namely, climatic change, related atmospheric phenomena, and their effects on human well-being. It then turns to a human dynamic, namely, the environmental consequences of the ways that medieval European social groups established and shifted their dietary habits.

Actively Changing "Nature" in Medieval Europe: Findings from Historical Climatology

Since the Pleistocene ice caps receded some eleven thousand years ago, the present period in planetary history, called the Holocene, has featured continual geophysical change and punctuated atmospheric change.[13] During the medieval millennium, specialists agree, Europe passed through three successive climatic phases: a cool period, roughly coincident with the transitional late antique and early medieval centuries; a "medieval warm period," during approximately the tenth through thirteenth centuries; and distinctly cooler overall temperatures which began in later medieval centuries and lasted into modern times. The last phase is now commonly, if imprecisely, called the Little Ice Age.[14] Although the glacial advances from which the Little Ice Age took its name seem to have occurred worldwide, few now deny that in regions without glaciers the same climatic oscillations had quite different effects.[15] Thinking about global shifts in weather patterns includes recognizing distinct regional manifestations.

The concurrent large-scale and long-term climatic trends affecting medieval Europe can be illustrated by looking at the results from three widely separated "field stations," located in Mediterranean Italy, continental central Europe, and the northwestern Atlantic. Scholars using data from these localities have deployed different methods to reconstruct historic patterns of temperature and precipitation with their geophysical effects.

Geoarchaeological examination of soils and sediments in Sicilian (38° N) and northern Italian (42–45° N) locations has established recurring climatic phases going back several millennia in the western Mediterranean basin.[16] Stratified sequences of soil types associated with well-dated human structures record shifts between warm-arid and cold-humid periods, each with mean annual temperatures deviating by up to 1.5°C from twentieth-century values. When the Mediterranean climate was relatively warm and dry, windborne sands covered existing soil surfaces and structures in the south, while in the north, reduced precipitation allowed organic soil to form; cold, humid periods brought waterborne erosion and deposition in the north and stabilized southern surfaces for soil formation. Transitional phases were typified by the strong seasonality of cool, wet winters and warm, dry summers now commonly thought characteristic of a "Mediterranean climate." Findings from these methods establish a "Dark Age Little Ice Age," when new alluvial sediments dating 500–750 were deposited in northern Italy, while Sicilian soil surfaces were more stable. A Mediterranean climate prevailed in centuries around the turn of the first millennium, but from roughly 1100 to 1270 southerners experienced desertification and sand deposits over existing soils, while stable soil formed in the north. More typically "Mediterranean" seasons again dominated during the fourteenth and fifteenth centuries, only to be succeeded after 1500 by northern erosion and southern stability under "Little Ice Age" conditions.

Central continental Europe provides a second exemplary place to observe the changing medieval climate, thanks to a systematic methodology pioneered by Swiss historian Christian Pfister and subsequently refined by various collaborators.[17] Briefly, Pfister's method combines contemporary verbal reports of meteorological and climatic events with such indirect (proxy) indicators as phenology[18] and palaeoscientific data sets from tree rings, lake bottom sediments, ice cores, and the like. Together, this evidence serves to identify and tabulate months, seasons, and years with temperatures or precipitation that were anomalously high or low, initially as compared to the simultaneous record of scientific observations in the same area. Once calibrated against the observational record and compared to present-day conditions, the statistical series is extended backward in time to the limits of critically established (i.e., correctly placed and dated) and relevant information.

Applying the Pfister method to sources from what is now Germany, Rüdiger Glaser found adequate records to cover 35 percent of the seasons in the eleventh century, 85 percent between 1100 and 1299, and 90 percent between 1300 and 1499.[19] Glaser's elaborate graphs sum up positive and negative seasonal anomalies at the decadal scale. The general findings indicate that medieval summers in central Europe were normally warmer and other seasons normally cooler and wetter than was typical of conditions there between ca. 1650 and 2000. But some periods distinctively departed from the medieval norm: while the second half of the thirteenth century had markedly warm spring, summer, and fall seasons, the next human generation in the first half of the fourteenth century experienced unusual year-round wetness, with both springs and falls abnormally cold. Later, in the second half of the fifteenth century, much of the year was again commonly both wetter and cooler than it had been in the immediate past or in most recent times. Two separate damp and chilly periods thus punctuated late medieval centuries in central Europe.

In the northwestern Atlantic, traces of past climates may be recovered from marine sediment cores taken around Iceland and ice cores from central Greenland. Both contain annual layers with trace isotopes of oxygen, whose ratio serves as a proxy for seasonal temperatures at the sea or ice surface, and intermittent marker layers of the ash deposited by dated volcanic eruptions. Joined to the verbal records of the Icelanders, these layers tell a single story.[20] This arctic region experienced relatively warm and stable temperatures throughout the eighth to twelfth centuries. The fourteenth century, however, was the coldest period in the entire seven hundred years of those records. Annual ice core data from central Greenland show deeply low temperatures during 1308–19, 1324–29, 1342–62 (when seasonal values show especially cold summers), and 1380–84. The Arctic sea ice pack, which formerly had remained well to the north, reached Iceland's shores in the winters of 1306, 1319–21, 1350, 1374, and normally throughout the 1400s.

All three groups of observations thus reveal long-term climatic change through three large phases with intervening transitions, but neither the decadal scale of timing nor the regional manifestations coincide. The discrepancies in part reflect the complex interplay of climatic variables in what is acknowledged to be a chaotic system. A medievalist ought not take issue with critically tested local and regional data sets, but rather should try to assemble a larger mosaic. It appears that the late medieval cooling trend propagated progressively eastward across Europe, with the British Isles and northwestern continent feeling its effects from the 1290s, the Mediterranean after 1320, and central Europe during midcentury decades. Apart from the greater storminess remarked below, cooling itself had different consequences in different geoclimatic regions. On Hungary's Pannonian plain, wetter conditions with a relatively low rate of evaporation reduced the frequency of dangerous summer droughts, while in eastern Mediterranean areas such as Crete, the Little Ice Age actually increased the risk of spring drought. The warm period in the Mediterranean coincided with greater seasonal differences in precipitation north of the Alps.[21]

Scholarship may not yet be ready thoroughly to explore and argue the effects on human subsistence and behavior of the long-term regional climatic synchronies so far identified during the Middle Ages. This will first require more regional documentation, using up-to-date methods in France, Britain, and Iberia, and then smaller-scale examination of economic and cultural phenomena in full awareness of climate. No medievalist can now, however, justify studying events in ignorance of the climatic and weather conditions in which they occurred. Although climate and weather certainly do not determine history, they set important framework conditions within which human activities—economic, cultural, political—had to take place.

Certain well-documented periods in the Middle Ages already vividly manifest climatic dynamism. The growing season of A.D. 536 initiated five to fifteen years of remarkably narrow tree rings in series taken from several different species of trees across Europe and elsewhere in the northern hemisphere.[22] In the late summer of 536, Cassiodorus, a Roman senator and the senior civil administrator of Italy, complained to a subordinate, "All of us are still observing, as it were, a blue-coloured sun; we marvel at bodies which cast no mid-day shadow, and at that strength of intensest heat reaching extreme and dull tepidity. And this has not happened in the momentary loss of an eclipse, but has assuredly been going on equally through almost the whole year." He described a "dry fog," which so reduced the sun's heat that crops had not grown properly and there was fear for the harvest.[23] The Byzantine historian Procopius, then present in either North Africa or Sicily, corroborates Cassiodorus's observations, and so do several other contemporary writers in Constantinople. Written records from China, Japan, and Korea and habitation sites and glacial ice cores from Peru confirm what today's atmospheric scientists would call a "dust veil event," which sharply cut solar radiation to the earth's surface.[24] Such events result from large volcanic eruptions or impacts of extraterrestrial objects, though attempts to identify a particular occurrence in 535 remain speculative.[25]

Storms and cold were widely reported during 536–37 and in years thereafter, followed by crop failures and a great drought in Italy in 539. From 542 a devastating epidemic spread across the Mediterranean and into western Europe. Given the few sources reporting on everyday life in the early Middle Ages, alleged consequences of environmental changes in politics and culture at some point become an untestable chain of contingencies.[26] Nevertheless, prolonged subsistence crises in the late 530s did bring unprecedented numbers of Slavs across the Danube into the Balkans and caused the Avar tribe to leave Mongolia for a generation-long migration to the west. At Europe's other extremity, the British resistance that had held Anglo-Saxon invaders for several decades in the eastern third of the island quietly disappeared in the middle of the sixth century. By shortly after A.D. 600, European populations had dwindled to a postclassical low point.

Comparable meteorological peculiarities also marked the entire early fourteenth century. Dendrochronological series suddenly show no growth of European oaks in 1318 and unusually low levels up to 1353. Human records identify peculiarly savage weather during 1314–17; this triggered mounting crop failures and epizootics among domestic livestock, which in turn laid the material conditions for widespread human famine and mortalities in 1315–19. A few well-studied areas show extraordinary storminess, marine inundations, and earthquakes in this and following decades.[27] The Great Aletsch Glacier, largest in the Alps, after surviving the previous five hundred years in a shrunken state, surged in a few decades to a near maximum extension.[28] The Arno at Florence flooded six times, mainly in autumn, between 1330 and 1362, but only two more times during the whole century.[29] The human disasters for which the fourteenth century is infamous occurred on a natural stage that was itself undergoing great perturbations.

Climatic change affects the resources on which human societies depend. Human communities respond within their own cultural parameters. It is necessary to think about threshold values: from ice floes, to frozen soil, to shortened growing seasons, to frost-killed olive trees; from arable smothered in sand, to silted rivers, to falling water tables and springtime droughts; to floods that damage urban and commercial infrastructure and delay plowing and harvesting. How much did it take to affect human livelihoods? Such concrete questions have first to be imagined, then asked, before they can be answered.

Consequences of medieval climatic change are, so far, most easily identified at high latitudes and high altitudes. Thus they notably involve effects attributed to the coming of the Little Ice Age. For example, Norse settlers arrived in Greenland from Iceland in the late tenth century with a subsistence strategy based on milk, meat, and fibre from cattle, sheep, and goats, which they overwintered indoors on summer-cut hay, supplemented by land-based capture of seals, seabirds, and caribou. They obtained foreign goods by exchange of walrus hides and ivory taken along the sea ice far to the north. Cooler conditions after 1250 reduced the time that livestock could feed on natural forage and the growth available for winter storage: animals had to be fed longer from smaller supplies of hay. Heavier sea ice shifted seasonal marine mammal concentrations away from the Norse settlements and closed access to far northern resources. Lacking the adaptation to fully marine hunting of their longtime Inuit neighbours, the Norse abandoned their more exposed Western Settlement between 1341 and 1362 and disappeared from their southern farms later in the fifteenth century.[30] Their contemporaries in Iceland suffered terrible deprivations, and many animals and humans died of starvation, but new adaptations replaced woollen homespun with dried codfish as a means of foreign exchange to support survival of their society.[31] Less dramatic but more widespread were such comparable adjustments as the retreat of oats cultivation in the Lammermuir hills of southeastern Scotland from 300 meters altitude in the thirteenth century to below 200 meters by the fifteenth.[32] Acute late medieval flare-ups of long-endemic malaria along low-lying North Sea coasts were plainly triggered by the arrival of chillier, damper conditions. Simultaneous upsurges of this disease also occurred in coastal Languedoc, southern mainland Italy, and Sicily, where they helped provoke abandonment of settlements on coastal lagoons. To what extent may these events also have resulted from a wetter climate, offering more suitable habitat to mosquito vectors?[33]

For the most part, medieval scholarship has yet to follow the leads provided by historical climatology and to investigate social and economic consequences from regional impacts of the warming conditions—Mediterranean desertification—in the central Middle Ages or of early medieval cooling.

One scientific climatologist, William Ruddiman, even argues that already in these preindustrial times, human behavior itself triggered feedback in the climatic system. He first hypothesizes that the observed long-term stability of Holocene temperatures resulted from balance between natural cooling and the warming effect of rising atmospheric carbon dioxide from human use of fire and early agricultural clearances of woodlands. He then proposes that temperature variations at century and decadal scales during the most recent two millennia (namely, the cold periods discussed above) were driven by regrowth of woodland after anomalous disease-caused human mortality peaks, including A.D. 200–600 and 1300–1400 in Eurasia, with intervening clearances bringing on the medieval warming.[34] Only medievalists have the skills and sufficiently precise data on the timing and location of human activity to test this hypothesis and its implications for the co-adaptation of humans and nature.[35]

Human Animals on the Trophic Pyramid: Environmental Consequences of Changing Medieval Dietary Cultures

Like natural forces—and independent of the hypothesized climatic effects of agricultural clearance and combustion—human culture was a strong driver of environmental change in medieval western Christendom. One good demonstration of this can be found by looking at dietary history from an ecological point of view.

Ecologists use the trophic pyramid (see fig. 1) as a graphic representation of the dependence of all terran life forms on the flow of energy from solar radiation: green plants ("primary producers") capture a small proportion of solar energy, and progressively lesser amounts ($\pm 10\%$) are consumed by plant-eating and then flesh-eating animals, culminating in the most rare "top carnivores." The actual relations of who eats whom in any given ecosystem are commonly portrayed as food chains intertwined in a food web. Changes in the specific identity or behavior of organisms in either the base or upper layers of an ecological food web can provoke minimal adjustments or trigger cascading change throughout the web, depending on how much the subject organism contributed to the structural stability of the web. Shifts involving "keystone species" send tremors up and down the pyramid.

Food consumption behaviors of the cultured animal, that is, of humans, fit smoothly into this analysis. Simply consider the well-known difference between the large mass of medieval Europeans, who obtained nearly all their calories from cereal grains, and the much smaller groups who, at certain times, locations, and social positions, consumed

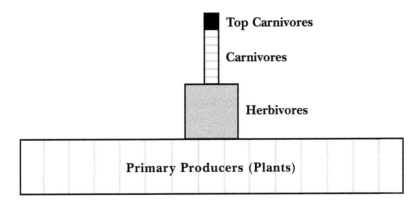

Figure 1. Trophic pyramid. Adapted from Ian G. Simmons, *The Ecology of Natural Resources*, 2nd ed. (London: Edward Arnold, 1982), 11.

large amounts of meat. Belgian zooarchaeologist Anton Ervynck initiated this line of reasoning with his 1992 article "Medieval Castles as Top-Predators of the Feudal System: An Archaeozoological Approach."[36] Extended application of the trophic perspective highlights the dynamic role of cultural forces in shaping medieval relations of *Homo in Natura*.

Archaeological studies of food remains identify barbarian and subsequent medieval elites in Europe north of the Alps as carnivores who commonly ate at the top of the pyramid.[37] In Flanders, Ervynck's original 1992 work found shared distinctive features at nineteen well-excavated castles. Inhabitants of these "top-predator" sites ate more animals, more diverse kinds of animals, and even more individuals of locally rare species than did people at other kinds of settlements. Their diets included significant numbers (though never a large proportion) of wild animals and many young domestic animals, whose tender and juicy flesh had been appropriated before the beast could be used for draft or fibre. Later research in the same region confirmed that rich kitchens of lay or clerical lords were "characterized by a high species variety, a selection for the best meat parts of a carcass, the abundance of the remains of young animals and the presence of game and rare and/or expensive species."[38] Flemish castle dwellers were especially fond of pork and the meat of wild animals, both mammals and birds, in great variety, while monastic tables carried little pork or game but many, often large or rare, fishes, fowl, and occasional beef or mutton. By the later Middle Ages, rich households in Flemish cities had also begun to eat much pig and wild game (including rabbit and birds).

Fifty-five medieval sites in northern France provide a comparably broad sample. Bones from the seigneurial residences consistently include more than double the share of wild animals found elsewhere, and, especially before the fourteenth century, those of pig were commonly more numerous than either cattle or sheep and goats.[39] Inhabitants of elite religious sites (monasteries) always ate less game than their secular counterparts, but by the late medieval centuries they ate just as much pork.[40]

Studies from a wider circle of regions replicate the pattern. While even very wealthy late Anglo-Saxon monasteries consumed less game than contemporary Anglo-Saxon lords, the monks liked their pork just about as well as did their untonsured cousins. These habits, including a greater diversity of wild fish and fowl at monastic settlements, continued into early Norman times. Pork subsequently lost favor, and few late medieval English sites have more remains of swine than of sheep.[41] Among Viking Age elites in Norway, Iceland, and even Greenland, the better the access to resources, the more pig and cattle were eaten.[42] So, too, in a comparative survey of twenty-eight central European sites about equally divided between Holstein and the Alpine foothills along the upper Rhine, pig bones were most common in earlier times and in fortified elite sites, but were progressively replaced by the bones of cattle, first, and especially, in urban settings.[43] Excavators of Ostrów Lednicki in Great Poland could track food waste from the castle and from a dependent borough, beginning in the mid-ninth century, when pigs were the leading source of meat for both. At the castle, a diminishing porcine dominance continued into the fifteenth century, but in the borough, beef consumption matched pork by the twelfth century and then surpassed it.[44] Likewise, bones from rare trophy European bison are uniquely associated in tenth- through thirteenth-century Hungary with "high status administrative centers."[45]

Other regional and social settings offer more or less sharp contrasts. Italian elite sites may contain more sheep, British and Hungarian more beef.[46] Flemish peasant settlements contain no game and only some cattle and sheep remains, and the same is true of urban sites predating the fourteenth century. Townspeople in medieval France ate less game and less pig but proportionately more cattle and sheep, while peasant settlements lack all game and provide the lowest share of pig and the highest of cattle and mutton. A research desideratum is to establish comparative patterns of meat consumption among medieval pastoral societies in the northern and western British Isles; in Spain, where long-distance seasonal movement of sheep was gaining importance; and in Scandinavia and Hungary (see below).

Diverse cultural considerations help contextualize elite carnivorous behavior. Meat eating certainly catered to elite taste, as it was generally thought the best food to sustain human muscles.[47] Pig flesh in particular received a top ranking from the Greco-Roman medical authority Galen (ca. 130–200) and, following him, more than a millennium of Christian physicians (Jewish and Muslim writers demurred).[48] For Germanic peoples, pork was most "meaty," perhaps for its fat content, and favored as promoting virility. Judging by food remains, it had been heavily consumed by those groups and by their opposite numbers at late imperial Roman military bases along the Rhine frontier.[49] The traditional monastic diet, however, originated in ascetics' purposeful rejection of elite Roman luxury—including pig eating—for a simplified version of the basic Mediterranean diet with little meat and some fish. The especially carnal pig was to be avoided, and not only for its taboo status in the Old Testament. But when it came to consumption by the laity, missionaries and clerics in the north, even the strict Irish authors of penitentials, did not ban pork. This selectivity was certainly related to the high status Celtic and Germanic cultures gave swine and their flesh; the Franks' *Salic Code*, for instance, ranked theft of pigs first among domestic animals.[50]

Early medieval pigs and game animals shared an equally close association with woodland, at a time when cattle in western Europe were completing a shift from the natural wooded-edge habitat of the ancestral *aurochs* to association with arable fields and hay meadows. Sheep commonly lived on open pasture or, in the setting of open-field agriculture, on the fallow and stubble.[51] Pigs were further set apart by having no productive function other than their meat. Hunting, of course, also displayed elite status, as, for clerics and laity alike, did the presence on table of rare, expensive, and showy animals.

The mid-fourteenth-century loss of perhaps a third of Europe's human population left survivors with higher wealth per capita, resulting in markedly greater meat consumption across the entire population.[52] Levels attained in the mid to late 1400s were not again reached in Europe before the late nineteenth and twentieth centuries. Annual rates of meat consumption came to fifty kilograms per head in Hamburg, forty-seven in south German towns, and twenty-six even at Carpentras in southeast France (Mediterranean consumption was typically less than in northern Europe). Cologne's mere thirty thousand human inhabitants in the 1490s slaughtered six to eight thousand head of cattle each year. Archaeozoological data from northern France[53] corresponds to inferences from the written record that average European consumption per head during 1350–1550 was three to four times greater than it had been during the eleventh to thirteenth centuries or would be after 1550.[54]

The power of culture-bearing top carnivores had environmental consequences in medieval Europe. Elite infatuation with the hunt motivated the creation of privileged jurisdictions, first royal "forests," then aristocratic "parks," where hunting took priority over other forms of land use. These developed into a prototypical special kind of managed landscape, not "wilderness" but more like wood pasture managed for game (deer, wild pig), resembling what we now call "parkland."[55] This behavior toward wild Nature had effects both protective—by retaining habitat and regulating consumptive use—and destructive—by modifying plant succession and extirpating rare trophy animals such as bear, wolf, and wild pig. Powerful medieval carnivores further undertook conscious environmental modification in their own interest. They introduced exotic animals (roe deer, rabbit, carp) to eat the meat. They shaped new kinds of anthropogenic landscapes at an unprecedented scale, for example, by constructing local clusters of purpose-built fish ponds that after the 1350s came to cover as much as three- to eight-hundred hectares each.

This setting provides historic examples of interactive co-adaptation, or the reciprocity of ecological change between Nature and humankind. Rising late medieval meat consumption favored beef over pork, notably among growing numbers of the better-off urban and rural commoners.[56] Greater western demand drove the development of large-scale long-distance movement of cattle to the west from new pastoral zones in southern Scandinavia and eastern Europe (Poland and Hungary). This economic innovation stressed producing ecosystems, selectively changed the physical character of the organisms produced (a new breed, the Hungarian Gray, appeared in the new export-oriented pastoralism), and shifted consumers' attitudes toward their own foodstuffs.[57] Late medieval consumption and trade in beef further reflected a long-term decline in woodland habitats of western Europe and a shift in land use priorities for that which remained. Stocks of game animals had been depleted; woods under coppice for fuel production barred semiwild hogs. Driven from their scavenging niches in the agro-pastoral landscape, surviving pigs were penned in farmyards and around towns to eat garbage.[58] But that very change in land use and resources serving medieval Europe's carnivores resulted equally from long-term environmental impacts of feeding behavior by the different and much more numerous group of medieval sociocultural herbivores.

While it was the few meat eaters who claimed power in medieval society, the whole civilization rested on a great biomass of herbivores, that is, the peasant consumers (and producers) of a predominantly cereal-based diet. Heavy human reliance on selected plant foods had commensurate environmental consequences.

The northern European dietary practices familiar to medievalists[59] depended historically on more or less arbitrary cultural decisions. An initial impetus may have been northern emulation of Mediterranean cultures, where "civilized" elites ate wheat bread. Frankish rulers actively promoted cereal consumption and cultivation[60] long before the pressures of general population growth would force people to capture the larger share of biological production available from eating more plants (see the trophic pyramid). Subsequently, of course, the number of poor peasants dependent on cheap calories grew in tandem with concentration on grain farming—to the extreme of the cereal monoculture established in the English Midlands under pressure of late Anglo-Saxon population increases[61] and comparable conditions in some parts of the continent. By around 1200, the inhabitants of central Europe were getting twice the proportion of their calories from grain as had their early medieval ancestors.[62]

Cultivation of cereal grasses propelled the great wave of woodland clearances which originated in the ninth-century Paris basin and the Rhineland, and by the thirteenth century had reached the Warwickshire Arden, the hills of Lorraine, and across the Odra river. Truly, as one Anglo-Saxon poet put it, the plowman was "the grey foe of the wood."[63] From sometime in the central Middle Ages, the long-term permanent conversion from woodland to arable distinguished northern European history from a Mediterranean pattern of successive depletion (overuse) of woodlands followed by their regrowth.[64] In what is now France, woodland dropped from 30 million hectares around 800 to 13 million by about 1300. Poland, where 16 percent of the land was arable in 1000, had, by the sixteenth century, 30 percent of it under cultivation; Prussia's 80 percent wooded cover had fallen to 60 percent by 1400. A mid-fourteenth-century Polish chronicler wrote that never before had so many new villages and towns been established where once only trees had grown.[65]

New arable fields created a new ecological niche, where humans captured a larger share of primary biological production. For the sake of those plants which medieval European herbivores wanted to eat, humans replaced old natural ecosystems with low productivity relative to high biomass of long-lived organisms (namely, trees) with an artificial agroecosystem containing a lower biomass of short-lived pioneer plants. High diversity of producing and consuming organisms gave way to low diversity, ideally a monoculture of annual cereal grasses, and to a short food chain, namely, some domestic herbivores and many, mainly plant-eating, humans.[66] Such ecosystems with low diversity of short-lived organisms and truncated food chains are characteristically unstable. They require continual inputs of energy to keep them at the pioneer stage, by preventing ecological succession of longer-lived woody plants.[67] Intensive cereal growing demanded more human labor and more natural plant nutrients throughout the annual production cycle than had earlier, more loosely mixed forms of agropastoralism.[68]

The new medieval agroecosystem disturbed the balance between plant cover, soil, and nutrients. Colonization on new lands and intensification on old lands initiated and accelerated large-scale soil erosion and deposition, which even led to the formation of new landscapes at some distance from the sites of human intervention. Broken vegetative cover, coupled with soil and nutrient loss, exposed large expanses of soil surface for seasonal removal, more often by water than by wind, and subsequent deposition. Bottom cores from the Lac d'Annecy (Savoy) show a sharp jump in sedimentation at the precise point in the thirteenth century when local monastic estates converted from woodland to arable exploitation.[69] In the Mediterranean basin, cumulative effects of small-scale clearances, drainage, and irrigation works greatly changed the hydrology. Coastal deposition and marsh formation in Sicily and southern peninsular Italy are evident from the twelfth century. Embayments along the Gulf of Lions, stable since Roman times, filled in or became lagoons during the Middle Ages.[70] In northern alluvial zones and estuaries the unintended changes induced by these human activities were abrupt and dramatic. Rates of alluviation in the upper Thames valley during the eleventh through thirteenth centuries surpassed all other postglacial periods. Soil erosion in Germany averaged less than five millimeters per annum for several millennia, but after woodland cover was reduced to a mere 10 percent of surface area, periods of extreme precipitation during 1313–19 caused the rate to peak at about five times the norm and again, in 1342, at fifteen times the norm.[71] As large-scale clearances at the top of the Vistula watershed changed its hydrographic balance during the thirteenth century, at the bottom of the watershed the silt and nutrients that had washed from new inland fields filled the historic bay between Gdańsk and Elbląg, created a chain of barrier islands,[72] and arguably contributed to the commercial failure of the herring schools that had for centuries spawned along the Pomeranian shore.

The larger ecological niche thus seized by human plant eaters changed conditions of life for all other organisms and strongly affected biodiversity. Loss of woody vegetation transformed terrestrial habitats, which altered carrying capacity for wild and domestic animals alike. Modern studies of bird life around England's last surviving open field village, Laxton in Nottinghamshire, found fewer species and half the density compared to nearby areas with mixed plant communities. The woodland species found to dominate the bird remains in archaeological sites around Madrid dating from the fifth through twelfth centuries had lost their importance by the later Middle Ages. Animals with other habitat preferences could be favored: populations of Great Bustard, a ground bird of open grasslands, increased in both England and Hungary.[73] Among domestic livestock, sheep and goats, well-adapted for living in interstices and edges of mixed arable landscapes, supplanted the woodland-dependent pig.

All kinds of humans fit into energy flow and thus have ecological place and consequences. Whether few powerful top carnivores or many herbivores, both have effects. Also in the Middle Ages human cultural changes provoked changes in surrounding Nature.

Medieval Europeans, though innocent of modern ecological concepts, were well aware of the complex and changing relationship between humans and their natural surroundings. Bernard Silvester and Paul Schneevogel alike saw *Homo* working *in Natura* to rear crops and livestock, to capture wild animals, and to extract fuel and minerals. They recognized some direct consequences of this human activity. Such contemporaries also acknowledged that *Natura* provided the sun, soil, and weather on which *Homo* had to depend. They knew that "temporales qui permutantur eventus" [temporal things subject to change] framed this relationship. Postmodern medievalists might profitably share the same historical consciousness and explore more fully the interactions between the mutually changing natural and cultural forces that helped to shape the lives of medieval people and their heirs.

Notes

This essay was first presented as a lecture in the series "Nature in the Middle Ages and Renaissance" at Ohio State University, Center for Medieval and Renaissance Studies, October 15, 2004, and then to the Historical Research Group at York University, December 2, 2004. I gratefully acknowledge advice and correction received from members of the audience at Ohio State, from colleagues at York, and from Verena Winiwarter, but retain sole responsibility for remaining errors and infelicities.

1. Bernard Silvestris [Bernardus Silvestris], *Bernardus Silvestris: Cosmographia*, ed. Peter Dronke (Leiden: Brill, 1978), and *The Cosmographia of Bernardus Silvestris*, trans. Winthrop Wetherbee (New York: Columbia University Press, 1973). Compare Brian Stock, *Myth and Science in the Twelfth Century: A Study of Bernard Silvester* (Princeton: Princeton University Press, 1972), and Haijo Jan

Westra, "Bernard Silvester," in *Dictionary of the Middle Ages*, ed. Joseph R. Strayer, vol. 2 (New York: Charles Scribner's Sons, 1983), 194–95.

    2. *Cosmographia: Microcosmos*, chaps. 10–12, ed. Dronke, 140–44; trans. Wetherbee, 113–16.

    3. Ibid., chap. 10, ll. 49–50, trans. Wetherbee, 114; ed. Dronke, 141: "Omnia subiiciat, terras regat, imperet orbi: / Primatem rebus pontificemque dedi."

    4. Ibid., "temporales qui permutantur eventus," ed. Dronke, 143.

    5. *Cosmographia: Microcosmos*, chap. 10, ll. 45–48, trans. Wetherbee, 114; ed. Dronke, 141: "Ut sua sint elementa volo: sibi ferveat ignis, / Sol niteat, tellus germinet, unda fluat. / Terra sibi fruges, pisces sibi nutriat unda, / Et sibi mons pecudes et sibi silva feras." For particulars on animals, plants, birds, and fishes, see *Megacosmos*, chap. 3.

    6. Bernard shared this view with such contemporary writers as Alan of Lille, Gilbert of Poitiers (de la Porée), and William of Conches, as described by David Herlihy, "Attitudes toward the Environment in Medieval Society," in *Historical Ecology: Essays on Environment and Social Change*, ed. Lester J. Bilsky (Port Washington, N.Y.: Kennikat Press, 1980), 100–116, with references to original scholarship by M.-D. Chenu.

    7. Paulus Niavis [Paul Schneevogel], *Iudicium iovis in valle amenitatis habitum ad quod mortalis homo a terra tractus propter montifodinas in monte niveo aliisque multis perfectas ac demum parricidii accusatus* (Leipzig: Martin Landsberg, n.d. [1492/95]), 16 fols. [British Library shelfmark I.A.11962]. Partially edited and translated by Paul Krenkel, *Iudicium Iovis oder Das Gericht der Götter über den Bergbau*, Freiberger Forschungshefte, Reihe Kultur und Technik, D3 (Berlin: Akademie, 1953), 13–38, with biographical notes on Schneevogel, 46–50. A loose summary from Krenkel appears in Johannes Grabmeyer, *Europa im späten Mittelalter, 1250–1500: Eine Kultur- und Mentalitätsgeschichte* (Darmstadt: Wissenschaftliche Buchgesellschaft, 2004), 135–36.

    8. Niavis, fol. A7r–v: "Terra autem que parentis nomen hominis vsurpat / et matris amorem in ipsus predicat abdit: atque occultat in intima sua / ut nouerce officium pocius quam genitricis habere videat Nec iupiter ipse nec quisque deorum dubitat / quin es quodlibet in hominum vtilitatem nascatur quamobrem necesse est iupiter id exquirere summaque diligentia investigare / pro tua nobis clementia elargitum / a terra invita occultatum."

    9. Niavis, fol. B8v: "Homines debere montes transfodere : metallifodinas perficere : agros colere studere mercature terramque offendere. scientiam abicere : plutonem inquietare . ac demum in riuulis aquarum venas metalli inquirere corpus vero eius a terra conglutiri : per vapores suffocari . vino inebriari : fame subici et quod optimum sit ignorare multa preterea alia pericula hominibus esse propria."

10. Herlihy, "Attitudes," likewise identified a medieval idea of "adversarial" relations between humans and nature, which he found in early medieval literature (i.e., *Beowulf*) and depictions of the forest in certain romances.

11. See exemplary statements by Daniel Botkin, *Discordant Harmonies: A New Ecology for the Twenty-First Century* (New York: Oxford University Press, 1990); William Balée, "Ecology, Historical," in *Encyclopedia of World Environmental History*, ed. Shepard Krech III, John R. McNeill, and Carolyn Merchant (New York and London: Routledge, 2004), 1:392–96; and Marina Fischer-Kowalski, "Ecology, Social," in ibid., 1:396–400.

12. As Theodore Steinberg sought to do for American history in "Down to Earth: Nature, Agency, and Power in History," *American Historical Review* 107 (2002): 798–820, and with more substantive detail in *Down to Earth: Nature's Role in American History* (New York: Oxford University Press, 2002).

13. William R. Dickinson, "Changing Times: The Holocene Legacy," *Environmental History* 5 (2000): 483–502, and more generally, N. Roberts, *The Holocene: An Environmental History* (Oxford: Blackwell, 1989).

14. Medievalists can survey the fast-developing field of historical climatology in Rudolf Brázdil, "Patterns of Climate in Central Europe since Viking Times," in *Climate Development and History of the North Atlantic Realm*, ed. G. Wefer et al. (Berlin and Heidelberg: Springer Verlag, 2002), 355–68, and Rudolf Brázdil, "Historical Climatology and Its Progress after 1990," in *People and Nature in Historical Perspective*, ed. József Laszlovszky and Péter Szabó (Budapest: Central European University, Department of Medieval Studies, and Archaeolingua, 2003), 197–228.

15. Jean M. Grove, *The Little Ice Age* (London: Methuen, 1988); Jean Grove, "The Century Time-Scale," in *Time-Scales and Environmental Change*, ed. T.S. Driver and G.P. Chapman (New York: Routledge, 1996), 39–87; J.M. Grove and R. Switsur, "Glacial Geological Evidence for the Medieval Warm Period," *Climatic Change* 30 (1994): 1–27; J.M. Grove, "The Onset of the Little Ice Age," in *History and Climate: Memories of the Future?*, ed. P.D. Jones, A.E.J. Ogilvie, T.D. Davies, and K.R. Briffa (New York: Kluwer, 2001), 153–86; J.M. Grove, "The Initiation of the Little Ice Age in Regions round the North Atlantic," *Climatic Change* 48 (2001): 53–82.

16. Franco Ortolani and Silvana Pagliuca, "Cyclical Climatic-Environmental Changes in the Mediterranean Area (2500 BP–Present Day)," *PAGES Past Global Changes News* 11:1 (April 2003): 15–17, which summarizes their extended work, *La variazioni climatiche storiche e la prevedibilità delle modificazioni relative all'effetto serra* (Rome: Asociazione Italiana Nucleare, 2001). Similar methods establish comparable climatic cycles keyed around humidity in an area of northern Tunisia barely 1° to the south of the Sicilian site: D. Faust, C. Zielhofer, F. Diaz del Olmo, and R.B. Escudero, "Fluvial Record of Late Pleis-

tocene and Holocene Geomorphic Change in Northern Tunisia—Global, Regional or Local Climatic Causes?" *PAGES Past Global Changes News* 13:1 (April 2005): 13–14.

17. Although Pfister began work in the 1970s, the first definitive publication of his methods was Christian Pfister, *Das Klima der Schweiz von 1525–1860 und seine Bedeutung in der Geschichte von Bevölkerung und Landwirtschaft*, 2nd ed. (Bern: Verlag Paul Haupt, 1985), vol. 1, with a good English-language summary at 161–64. A third, slightly revised edition came out in 1988. For further methodological development see Christian Pfister, "Variations in the Spring-Summer Climate of Central Europe from the High Middle Ages to 1850," in *Long and Short Term Variability of Climate*, ed. H. Wanner and U. Siegenthaler (Berlin, Heidelberg, New York: Springer Verlag, 1988), 57–82; Werner Schwarz-Zanetti, Christian Pfister, Gabriela Schwarz-Zanetti, and Hannes Schüle, "The EURO-CLIMHIST Data Base—A Tool for Reconstructing the Climate of Europe in the Pre-instrumental Period from High Resolution Proxy Data," in *European Climate Reconstructed from Documentary Data: Methods and Results*, ed. Burkhard Frenzel (Stuttgart: Fischer Verlag, 1992), 193–210; Joël Guiot, "The Combination of Historical Documents and Biological Data in the Reconstruction of Climate Variations in Space and Time," in ibid., 94–104; various authors' contributions to *Climatic Trends and Anomalies in Europe 1675–1715: High Resolution Spatio-Temporal Reconstructions from Direct Meteorological Observations and Proxy Data: Methods and Results*, ed. Burkhard Frenzel (Stuttgart and New York: G. Fischer, 1994); and Christian Pfister, *Wetternachhersage: 500 Jahre Klimavariationen und Naturkatastrophen (1496–1995)* (Bern: Paul Haupt, 1999).

18. Compare, for example, Isabelle Chuine, Pascal Yiou, Nicolas Viovy, Bernard Seguin, Valérie Daux, and Emmanuel Le Roy Ladurie, "Historical Phenology: Grape Ripening As a Past Climate Indicator," *Nature* 432 (18 November 2004): 289–90.

19. Rüdiger Glaser, *Klimageschichte Mitteleuropas: 1000 Jahre Wetter, Klima, Katastrophen* (Darmstadt: Wissenschaftliche Buchgesellschaft, 2001).

20. A. E. Ogilvie, "Climatic Changes in Iceland A.D. ca. 865 to 1598," *Acta Archaeologica* 61 (1990): 233–51; L. K. Barlow, J. C. Rogers, M. C. Serreze, and R. G. Barry, "Aspects of Climate Variability in the North Atlantic Sector: Discussion and Relation to the Greenland Ice Sheet Project 2 High-resolution Isotopic Signal," *Journal of Geophysical Research* 102 (C12):26 (1997): 333–44; Thomas H. McGovern, "The Demise of Norse Greenland," in *Vikings: The North Atlantic Saga*, ed. William W. Fitzhugh and Elizabeth Ward (Washington, D.C.: Smithsonian Institution Press, 2000), 327–39; Astrid E. J. Ogilvie and Thomas H. McGovern, "Sagas and Science: Climate and Human Impacts in the North Atlantic," in ibid., 385–93; J. Eiriksson, K. I. Knudsen, H. Haflidason,

and J. Heinemeier, "Chronology of Late Holocene Climatic Events in the Northern North Atlantic Based on AMS C-14 Dates and Tephra Markers from the Volcano Hekla, Iceland," *Journal of Quaternary Science* 15 (2000): 573–80.

21. A. T. Grove and Oliver Rackham, *The Nature of Mediterranean Europe: An Ecological History* (New Haven: Yale University Press, 2001), 130–40.

22. M. G. L. Baillie, "Dendrochronology Provides an Independent Background for Studies of the Human Past," in *L'uomo e la foresta secc. XIII–XVIII*, ed. Simonetta Cavaciocchi, Instituto Internazionale di Storia Economica "F. Datini," Prato, Serie II—Atti delle "Settimane di Studi" e altri Convegni, vol. 27 (Florence: Monnier, 1996), 99–119, and "Putting Abrupt Environmental Change Back into Human History," in *Environments and Historical Change: The Linacre Lectures 1998*, ed. Paul Slack (Oxford: Oxford University Press, 1999), 46–75.

23. "Cernimus adhuc cuncti quasi uenetum solem: miramur media die umbras corpora non habere et uigorem illum fortissimi caloris usque ad extremi teporis, inertiam peruenisse, quod non eclipsis momentaneo defectu, sed totius paene anni agi nihilominus constat excursu." Cassiodorus, *Variarum libri duodecim*, 12:25, ed. A. J. Fridh, Corpus Christianorum, Series Latina, vol. 96 (Turnhout: Brepols, 1973), 492–94; ed. and trans. S. J. B. Barnish, *Cassiodorus: Variae* (Liverpool and Philadelphia: Liverpool University Press, 1992), 179–81.

24. David Keys, *Catastrophe* (London: Century, 1999); J. D. Gunn, ed., *The Years Without Summer: Tracing AD 536 and Its Aftermath*, BAR International Series 872 (Oxford: Archaeopress, 2000).

25. Several well-dated "dry fog" events in the central Mediterranean during the last phase of the Little Ice Age (1780s–1820s) are associated with well-documented major volcanic events (Grove and Rackham, *Nature of Mediterranean Europe*, 137–38).

26. As drawn out in Neville Brown, *History and Climatic Change: A Eurocentric Perspective* (New York: Routledge, 2001), 91–93.

27. Mark Bailey, "*Per impetum Maris*, Natural Disaster and Economic Decline in Eastern England, 1275–1350," in *Before the Black Death: Studies in the "Crisis" of the Early Fourteenth Century*, ed. B. M. S. Campbell (Manchester: Manchester University Press, 1991), 184–208; Michel Morineau, "Cataclysmes et calamités naturelles aux Pays-Bas septentrionaux XIe–XVIIIe siècles: Le travaille de la planète et la rétorsion des hommes," in *Les catastrophes naturelles dans l'Europe médiévale et moderne*, ed. B. Bennassar, Actes des XVes Journées Internationales d'Histoire de l'Abbaye de Flaran, 10, 11, 12 Septembre 1993 (Toulouse: Presses Universitaires du Mirail, 1996), 42–59. On the *Grosse Mandrank* flood of 1362 in East Friesland, see Glaser, *Klimageschichte Mitteleuropas*, 89.

28. Wilfried Haberli and Hanspeter Holzhauser, "Alpine Glacier Mass Changes during the Past Two Millennia," *PAGES Past Global Changes News* 11:1 (April 2003): 13–15.

29. Grove and Rackham, *Nature of Mediterranean Europe*, 135–36.

30. Thomas H. McGovern et al., "Northern Islands, Human Error, and Environmental Degradation: A View of Social and Ecological Change in the Medieval North Atlantic," *Human Ecology* 16 (1988): 225–70; Ogilvie, "Climatic Changes in Iceland"; R. C. Buckland et al., "Bioarchaeological and Climatological Evidence for the Fate of the Norse Farmers in Medieval Greenland," *Antiquity* 70 (1996): 88–96; McGovern, "Demise of Norse Greenland"; Ogilvie and McGovern, "Sagas and Science."

31. Kirsten Hastrup, *Culture and History in Medieval Iceland: An Anthropological Analysis of Structure and Change* (Oxford: Clarendon, 1985) and *Nature and Policy in Iceland 1400–1800: An Anthropological Analysis of History and Mentality* (Oxford: Clarendon, 1990); Gísli Pálsson, "The Idea of Fish: Land and Sea in the Icelandic World View," in *Signifying Animals: Human Meaning in the Natural World*, ed. R. Willis (London: One World Archaeology, 1990), 119–33.

32. J. M. Grove, "Century Time-Scale," 62–63, using data from M. L Parry, "Secular Climatic Change and Marginal Land," *Transactions of the Institute of British Geographers* 64 (1975): 1–13. Grove also gives other examples of lower altitude limits for agriculture in the late medieval British Isles.

33. Otto S. Knottnerus, "Malaria around the North Sea: A Survey," in *Climate Development and History of the North Atlantic Realm*, ed. G. Wefer, W. Berger, K.-E. Wehre, and E. Jansen (Berlin, Heidelberg, New York: Springer Verlag, 2002), 339–53. Incidental references to post-twelfth-century resurgence of malaria in specific regions around the western Mediterranean appear in Henri Bresc, "La pêche dans l'espace économique normand," in *Terra e uomini nel Mezzogiorno normanno-svevo*, Atti delle settime giornate normanno-sveve, Bari 15–17 ottobre 1985, ed. G. Musca (Bari: Centro di studi normanno-svevi Università degli Studi di Bari, 1987), 274–75; Henri Bresc, "Pêche et habitat en méditerranée occidentale," in *Castrum 7: Zones côtières littorales dans le monde méditerranéen au Moyen Âge: Defense, peuplement, mise en valeur*, Actes du colloque international organisé par l'École française de Rome et la Casa de Velázquez, en collaboration avec le Collège de France et le Centre interuniversitaire d'histoire et d'archéologie médiévales (UMR 5648–Université Lyon II–C.N.R.S.–E.H.E.S.S.). Rome, 23–26 octobre 1996, ed. Jean-Marie Martin (Rome-Madrid: École française de Rome, Casa de Velázquez, 2001), 532–38; and Monique Bourin-Derruau, Daniel Le Blévec, Claude Raynaud, and Laurent Schneider, "Le littoral languedocien au Moyen Âge," in ibid., 416–18.

34. William F. Ruddiman, "The Anthropogenic Era Began Thousands of Years Ago," *Climatic Change* 61 (2003): 261–93; "Early Anthropogenic Overprints on Holocene Climate," *PAGES Past Global Changes News* 12:1 (April 2004): 18–19; and "How Did Humans First Alter Global Climate?" *Scientific American* (March 2005): 46–53.

35. This is simply not the place to explore other topics of exciting current research on dynamic natural forces during the Middle Ages, whether the historical epidemiology of human (and animal) pathogens implicated by Ruddiman, or such geophysical phenomena as earthquakes, vulcanism, and landslides.

36. *Chateau Gaillard* 15 (1992): 151–59. The limited space of the present essay also precludes examining here the environmental consequences of other dietary differences, for instance, drinking wine or drinking beer, with their distinctive material and energy demands.

37. Elite carnivorism—though not its ecological implications—is also inferred from verbal sources by Antoni Riera-Melis, "Society, Food and Feudalism," in *Food: A Culinary History from Antiquity to the Present*, ed. Jean-Louis Flandrin and Massimo Montanari, with Albert Sonnenfeld, and trans. Clarissa Botsford et al. [revised English edition of their *Histoire de l'alimentation*] (New York: Columbia University Press, 1999), 251–67; Allen J. Grieco, "Food and Social Classes in Late Medieval and Renaissance Italy," in ibid., 302–12.

38. Anton Ervynck, "Following the Rule? Fish and Meat Consumption in Monastic Communities in Flanders (Belgium)," in *Environment and Subsistence in Medieval Europe*, ed. Guy De Boe and Frans Verhaeghe, Papers of the "Medieval Europe Brugge 1997" Conference, vol. 9, I.A.P. Rapporten (Zellik, Belgium: Instituut voor het Archeologisch Patrimonium, 1997), 78; Anton Ervynck, "*Orant, pugnant, laborant:* The Diet of the Three Orders in the Feudal Society of Medieval North-western Europe," in *Behaviour behind Bones: The Zooarchaeology of Ritual, Religion, Status and Identity*, ed. Sharyn Jones O'Day, Wim van Neer, and Anton Ervynck, Proceedings of the 9th Conference of the International Council of Archaeozoology, Durham, August 2002 (Oxford: Oxbow Books, 2004), 215–23.

39. Because sheep and goat bones can rarely be separated in an archaeological context, they are commonly treated together as "ovicaprids."

40. Benoît Clavel, *L'Animal dans l'alimentation médiévale et moderne en France du nord (XIIe–XVIIe siècles)*, Revue Archéologique de Picardie, No. Spécial 19 (n.p.: CRAVO, 2001), 98–114.

41. Alan Hardy, Anne Dodd, Graham D. Keevil, et al., *Ælfric's Abbey: Excavations at Eynsham Abbey, Oxfordshire, 1989–92*, Oxford Archaeology, Thames Valley Landscapes, 16 (Oxford: Oxford University School of Archaeology for Oxford Archaeology, 2003), 402–6 and works there cited.

42. McGovern "Demise of Norse Greenland," 332.

43. Heide Hüster-Plogmann and André Rehazek, "1000 Years (6th to 16th Century) of Economic Life in the Heart of Europe: Common and Distinct Trends in Cattle Economy of the Baltic Sea Region and the Swiss Region of the Alpine Forelands," *Archaeofauna* 8 (1999): 123–33.

44. Daniel Makowiecki, *Hodowla oraz użytkowanie zwierząt na Ostrowie Lednickim w średniowieczu: Studium archeozoologiczne*, Bibliotka studiów lednickich, vol. 6 (Poznań: Muzeum Pierwszych Piastów na Lednicy, 2001), 51–85. The age of slaughtered pigs indicates more were killed in spring than in supposedly traditional autumn.

45. László Bartosiewicz, "People and Animals: The Archaeozoologist's Perspective," in *People and Nature*, ed. Laszlovszky and Szabó, 31.

46. Frédérique Audoin-Rouzeau, "Elevage et alimentation dans l'espace européen au Moyen Âge: Cartographie des ossements animaux," in *Milieux naturels, espaces sociaux: Études offertes à Robert Delort*, ed. Elisabeth Mornet and Franco Morenzoni (Paris: Publications de la Sorbonne, 1997), 143–59.

47. Massimo Montanari, "From the Late Classical Period to the Early Middle Ages," in *Food*, ed. Flandrin and Montanari, 179. In the words of encyclopedist Bartholomeus Anglicus, following numerous medical authorities, "Carnes autem animalium prout sunt ordinatae ad humanum esum, variantur secundum diuersam praeparationem." Bartholomaeus Anglicus, *De rerum Proprietatibus* (Frankofurti: apud Wolfgangum Richterum, 1601 [facsimile reprint, Frankfurt: Minerva, 1964]), Lib. 18, cap. 1, pp. 984–85.

48. Ken Albala, *Eating Right in the Renaissance* (Berkeley: University of California Press, 2001), 77.

49. Montanari, "From the Late Classical Period," 167–69; Audoin-Rouzeau, "Elevage," 145.

50. Bernadette Filotas, *Pagan Survivals, Superstitions and Popular Cultures in Early Medieval Pastoral Literature* (Toronto: Pontifical Institute of Mediaeval Studies, 2005), 350–51.

51. See, for example, Tom Williamson, *Shaping Medieval Landscapes: Settlement, Society, Environment* (Macclesfield, Cheshire: Windgather Press, 2003), 22–23, 79–80, and 132–38, and Richard C. Hoffmann, *Land, Liberties, and Lordship in a Late Medieval Countryside: Agrarian Structures and Change in the Duchy of Wrocław* (Philadelphia: University of Pennsylvania Press, 1989), 99–100.

52. Grieco, "Food and Social Classes," 309–11; Vern Bullough and Cameron Campbell, "Female Longevity and Diet in the Later Middle Ages," *Speculum* 55 (1980): 317–25. Audoin-Rouzeau, "Elevage," 145, also notes a brief resurgence of pig remains.

53. Clavel, *L'Animal*, 98–102.

54. Diedrich Saalfeld, "Der Boden als Objekt der Aneigung," in *Von der Angst zur Ausbeutung: Umwelterfahrung zwischen Mittelalter und Neuzeit*, ed. Ernst Schubert and Bernd Herrmann (Frankfurt: Fischer Verlag, 1994), 75.

55. John Cummins, "*Veneurs s'en vont en Paradis*: Medieval Hunting and the 'Natural' Landscape," in *Inventing Medieval Landscapes: Senses of Place in Western Europe*, ed. John Howe and Michael Wolfe (Gainesville: University Press of

Florida, 2002), 33–56; Oliver Rackham, *The History of the Countryside* (London: Dent, 1986), 119–52, and *Trees and Woodland in the British Landscape*, rev. ed. (London: Dent, 1990), 151–83.

56. Clavel, *L'Animal*, 98–102, and, more generally, Audoin-Rouzeau, "Elevage."

57. Richard C. Hoffmann, "Frontier Foods for Late Medieval Consumers: Culture, Economy, Ecology," *Environment and History* 7 (2001): 137–40.

58. Ervynck, "*Orant, pugnant, laborant*," 221; Florian Ruhland, "Schweinehaltung in und vor der Stadt," in *Nürnberg: Archäologie und Kulturgeschichte*, ed. Birgit Friedel and Claudia Frieser (Nürnberg: Verlag Dr. Faustus, 1999), 319–25; Kathleen Biddick, *The Other Economy: Pastoral Husbandry on a Medieval Estate* (Berkeley and Los Angeles: University of California Press, 1989), 132; Hardy et al., *Ælfric's Abbey*, 404–6.

59. Kathy L. Pearson, "Nutrition and the Early-Medieval Diet," *Speculum* 72 (1997): 1–32; Saalfeld, "Boden," 75; Christopher Dyer, "English Diet in the Later Middle Ages," in *Social Relations and Ideas*, ed. T.H. Aston et al. (Cambridge: Cambridge University Press, 1983), 191–216, and "Changes in Diet in the Late Middle Ages: The Case of Harvest Workers," *The Agricultural History Review* 36 (1988): 21–38; Riera-Melis, "Society, Food and Feudalism"; Grieco, "Food and Social Classes."

60. Adriaan Verhulst, *The Carolingian Economy* (Cambridge: Cambridge University Press, 2002), 63–66; Christoph Sonnlechner, "The Establishment of New Units of Production in Carolingian Times: Making Early Medieval Sources Relevant for Environmental History," *Viator* 35 (2004): 21–48; Georges Comet, "Les céréales du Bas-Empire au Moyen Âge," in *The Making of Feudal Agricultures?*, ed. Miquel Barceló and François Siguat (Leiden: Brill, 2004), 131–76.

61. Williamson, *Shaping Medieval Landscapes*, 28–90.

62. Saalfeld, "Boden," 75.

63. "Geonge swa me wisað / har holtes feond," Riddle 21, ll. 2b–3a, in *The Exeter Book*, ed. George P. Krapp and Elliott V. K. Dobbie (New York: Columbia University Press, 1936), 191.

64. Grove and Rackham, *Nature of Mediterranean Europe*, 188–89.

65. Philippe Leveau, "L'archéologie des paysages et les époques historiques: Les grands aménagements agraires et leur signature dans le paysage (anthropisation des milieux et complexité des sociétés)," in *Milieux Naturels: Espaces Sociaux: Études offertes à Robert Delort* (Paris: Publications de la Sorbonne, 1997), ed. Élisabeth Mornet and Franco Morenzoni, 71–84; J. Piskorski, "The Historiography of the So-called 'East Colonisation' and the Current State of Research," in *The Man of Many Devices, Who Wandered Full Many Ways . . . : Festschrift in Honor of János M. Bak*, ed. Balázs Nagy and Marcell Sebők (Budapest: Central European University Press, 1999), 658–59, and works there cited.

66. S.W. Green, "The Agricultural Colonization of Temperate Forest Habitats: An Ecological Model," in *The Frontier: Comparative Studies*, vol. 2, ed. W.W. Savage, Jr., and S. Thompson (Norman: University of Oklahoma Press, 1979), 69–103; U. Emanuelsson, "The Relationship of Different Agricultural Systems to the Forest and Woodlands of Europe," in *Human Influence on Forest Ecosystems Development in Europe*, ed. F. Salbitano (Bologna: [n.p.], 1989), 169–78.

67. William S. Cooter, "Ecological Dimensions of Medieval Agrarian Systems," *Agricultural History* 52 (1978): 458–77 (with responses by R.S. Loomis and J.A. Raftis, 478–87); H.S.A. Fox, "Some Ecological Dimensions of Medieval Field Systems," in *Archaeological Approaches to Medieval Europe*, ed. Kathleen Biddick (Kalamazoo, Mich.: Medieval Institute Publications, 1984), 119–58; William S. Cooter, "Environmental, Ecological, and Agricultural Systems: Approaches to Simulation Modeling Applications for Medieval Temperate Europe," in ibid., 159–70.

68. See, for example, David Postles, "Cleaning the Medieval Arable," *The Agricultural History Review* 37 (1989): 130–43; J. Pretty, "Sustainable Agriculture in the Middle Ages on the English Manor," *The Agricultural History Review* 38 (1990): 1–19; R.S. Shiel, "Improving Soil Productivity in the Pre-fertilizer Era," in *Land, Labour, and Livestock: Historical Studies in European Agricultural Productivity*, ed. Bruce M.S. Campbell and Mark Overton (Manchester: Manchester University Press, 1991), 51–77; Bruce M.S. Campbell and Mark Overton, "A New Perspective on Medieval and Early Modern Agriculture: Six Centuries of Norfolk Farming, c.1250–c.1850," *Past & Present* 141 (1993): 38–105.

69. F. Oldfield and R.L. Clark, "Environmental History—The Environmental Evidence," in *The Silent Countdown: Essays in European Environmental History*, ed. Peter Brimblecombe and Christian Pfister (Berlin, Heidelberg, New York: Springer Verlag, 1990), 152–55; D.S. Crook, D.J. Siddle, J.A. Dearing, and R. Thompson, "Human Impact on the Environment in the Annecy Petit Lac Catchment, Haute Savoie: A Documentary Approach," *Environment and History* 10 (2004): 247–84.

70. Bresc, "La pêche," 274–75; Bourin-Derruau et al., "Littoral languedocien," 345–57.

71. Hans-Rudolf Bork and Gabrielle Schmidtchen, "Boden: Entwicklung, Zerstörung und Schutzbedarf in Deutschland," *Geographische Rundschau* 53:5 (May 2001): 4–9.

72. Teresa Dunin-Wąsowicz, "Environnement et habitat: La rupture d'équilibre du XIIIe siècle dans la Grande Plaine Européene," *Annales E.S.C.* 35:5 (Septembre–Octobre 1980), 1026–45, and "Natural Environment and Human Settlement over the Central European Lowland in the 13th Century," in *Silent*

Countdown, ed. Brimblecombe and Pfister, 92–105; J. Filuk, "Biologiczno-rybacka charakterystyka ichtiofauny zalewu wiślanego na tle badań paleo-ichtiologicznych, historycznych i wspólczesnych," *Pomorania antiqua* 2 (1968): 146–48.

73. Eric L. Jones, "The Bird Pests of British Agriculture in Recent Centuries," *The Agricultural History Review* 20 (1972): 118; Arturo Morales Muñiz and Dolores Carmen Morales Muñiz, "¿De quién es este ciervo?: Algunas consideraciones en torno a la fauna cinegética de la España medieval," in *El medio natural en la España medieval: Actas del I Congreso sobre ecohistoria e historia medieval*, ed. Julián Clemente Ramos (Cáceres: Universidad de Extremadura, 2001), 397–99 and 404–6; Erika Gál, "Adaptation of Different Bird Species to Human Environments," in *People and Nature*, ed. Laszlovszky and Szabó, 121–38.

CHAPTER TWO

Inventing with Animals in the Middle Ages

Jeffrey Jerome Cohen

The last few years have seen an outpouring of scholarship engaged in rethinking the interrelation of humans and animals. This boundary-challenging work mainly explores the precariousness of that divide we imagine separating us from other mammals.[1] My preoccupation with the animals of the Middle Ages is spurred in part by this critical efflorescence, with its bracing challenge to the supposed solitariness of species and identity. My interest also derives from the fact that I have two children. At the age of nineteen months, my daughter Katherine self-identified more strongly with monkeys than with *homo sapiens*. Her nursery a rainbow-colored menagerie, her picture books bursting with fantastic zoos, she resides in a hyperactive world of fauna. As the Disney megacorporation realized long ago, and as Katherine is realizing now, animals teach children how to become human. They also provide a temporary escape from that burden. Over the past few years I have been reading my young son Alexander a nightly installment of Brian Jacques's *Redwall* novels. The books feature abbesses and armored warriors, perilous weapons, feasts, foundlings, tapestries, stained glass — the characters and substance of the medieval world. Yet the actors in

39

these tales are mice, shrews, hawks, stoats, badgers, and weasels. Although the *Redwall* novels create an imaginary geography where beasts enact medieval dramas, through their speech and their actions it is clear that these animals inhabit a nostalgic fantasy of the British Empire. Jacques's medieval beasts open an imaginative space where the problems of a complex present can be simplified, and where good and evil are as self-evident as the animal skin one dwells inside. For all its talking creatures, the *Redwall* books are ultimately populated by humans. Animals are the vehicles through which desires for a differently configured (if obdurately anthropocentric) world are expressed.

Animals similarly offered "possible bodies" to the dreamers of the Middle Ages, forms both dynamic and disruptive through which might be dreamt alternate and even inhuman worlds.[2] In animal flesh could be realized some potentialities for identity that might escape the constricting limits of contemporary race, gender, or sexuality. Animals were fantasy bodies through which denied enjoyments might be experienced and foreclosed potential opened to exploration. For the most part, the purpose of inventing with animals was to yield more possibilities for humans, and therefore it proceeded only from a historical and rather limited point of view. Yet medieval authors and artists could, at least implicitly, approach the animal nonanthropomorphically. Though never likely, it was nonetheless sometimes possible to see in the beast more than a mere semblance of the human, not some lifeless allegory or a thing so nonhuman as to be wholly other. At times it was possible to grant to the animal its enduring status as intimate alien, as an intractable and ahistorical melding of the familiar and the strange.

Intimacy and Fantasy

Perhaps the most powerful moment of Chaucer's Knight's Tale occurs when the lovelorn warrior Palamon, doomed to perpetual imprisonment, rails against the gods for their cruelty. He wonders if people are any more valuable to the divinities than "the sheep that rouketh [cowers] in the folde."[3] Humans must keep their desires under control, Palamon observes, while "a beest may al his lust fulfille" (I.1318).

To make matters worse, even those who lead miserable lives are punished in the afterworld for their sins, while "whan a beest is deed he hath no peyne" (I.1319). Palamon imagines that to be an animal is to live in freedom, to indulge desire with impunity. Humans gain nothing but torment from the souls that differentiate them from beasts, and from the sense of time that makes them all too aware of the brevity of their mundane span. Chaucer was too pious an author to argue that humans would be better off without the judgment of God hovering over their heads, but he frequently employed animals as vehicles through which complaints against the difficulties of existence as a constrained being could be voiced. The Canterbury tale with the most sex in it is, after all, the Nun's Priest's Tale, where twenty acts of copulation occur between a rooster and his favorite wife (albeit in some very swift lines, VII.3177–78).

That Chaucer explored human limits through animal bodies is not surprising. Animals were in the Middle Ages what might be called proximate strangers, creatures intimate with humans and human-like in many ways, but forever set apart from the men, women, and children whose lives were inseparably bound to their own. Cohabitation with domesticated animals was an everyday fact of medieval life, as evident in cities and villages as on farms. Dogs, cats, pigs, horses, sheep, and birds could bear personalizing names. Unlike contemporary pets, however, medieval animals seldom carried the same appellations used among humans. No one pretended that a dog was a family member arrested in permanent childhood, as we do today. Chaucerian canines like Colle (coal) and Gerland (garland) in the Nun's Priest's Tale take their designations from coloration and markings, not from common English names.[4] The intimacy of animal and human extended far, for medieval peoples clothed themselves in hides and furs; used leather as we do plastic; built their dietary laws around the consumption of meat; and of course employed the reproductive fluids (i.e., milk) of sheep, goats, mares, and cows to create food and drink.[5]

Foxes, fowl, otters, deer, and badgers were nature's bounty in the countryside, to be hunted for meat, entertainment, or fur. Captive animals such as bears and boars featured in sport and were exhibited as curiosities. According to William of Malmesbury, England's Henry I was an avid collector of animals from foreign lands:

> Henry took a passionate delight in the marvels of other countries . . . asking foreign kings to send him animals not found in England—lions, leopards, lynxes, camels—and he had a park called Woodstock in which he kept his pets of this description.[6]

William himself saw an African porcupine there and marveled at its spikes. Woodstock was filled with wild beasts transformed into displays of a king's power and cosmopolitanism, an ebullient commingling of the indigenous and the exotic. Interestingly, Woodstock was also the rural estate where the Norman, francophone Henry maintained an English mistress and numerous illegitimate children. William of Malmesbury said of Henry that he was ever "chasing after whores." Orderic Vitalis was blunter, accusing Henry of "brainlessly rutting like a mule."[7] Though no contemporary writer connected the foreign menagerie Henry assembled at Woodstock with the extended mixed-race ménage he created there, it is difficult not to link the two through sexualized animal bodies, as Orderic does implicitly.

Although medieval animals could at times be objects of affection, most—whether domesticated, captured, or hunted—were valued according to their human use. Medieval writers saw the subordination of animals as a natural fact for which they could invoke classical, biblical, and patristic precedent. Aristotle had classified humans as animals but had never questioned innate human superiority over beasts lacking in reason. The creation story in Genesis had given man domination over the living creatures of the world, stressing the superiority of Adam over the fauna he had named.[8] Animal meaning, in other words, proceeded anthropocentrically. The idea of human dominance therefore had ethical and theological repercussions; it downplayed the potential autonomy of the animal by assuming a primordial and utter difference.[9]

From Aesop's fables in the classical period to Marie de France's *Le Dit d'Ésope* in the medieval, however, animals served well as creatures from whose actions human lessons might be extracted. "The Lord created different creatures," wrote Thomas of Chobham, "with different natures not only for the sustenance of men, but for their instruction." From contemplation of animals, continued this English theologian of the thirteenth century, we can discover "what may be useful to the soul."[10] Alan of Lille similarly argued that every animal functions as *liber, pictura, speculum* (book, image, mirror).¹¹ Reading the animal is a pedagogical process of glimpsing the human where the bestial used to be, of looking at what is in fact a combination of difference and similarity but comprehending only ourselves. Following Alan's lead, Nona C. Flores in *Animals in the Middle Ages* argues that medieval animals should be interpreted "not as literal living organisms—food, prey, possessions, or companions to man—but as symbols, ideas, or images" (ix). Animals, like nature more generally, become not beings or phenomena that exist for their own sake but vehicles through which human meanings are expressed. As metaphors and as allegories animals become pedagogical bodies, instructing us in how to sustain an identity that is not always easy to support. Thus Aesop's fable of the fox and the grapes teaches its readers a rather banal moral about the human failing of greed. His story of the artist and the lion, quoted so enthusiastically by Chaucer's Wife of Bath, is a lesson in perspective that would seem to suggest the world need not be viewed solely through human eyes (a lion tells a sculptor that, had he carved the victory statue, the beast would be triumphing over the hero). Yet as Alisoun's deployment of the story in her Prologue to mark gendered ways of narrating history suggests, the fable is more about the multiplicity of human points of view than an argument for the possibility of nonhuman points of view.

Not every animal is a human in a beast's body, even when the animal is deployed as a meditation on the burdens of human identity. In the Old English poem "The Wanderer," the forlorn narrator voices his isolation through the bodies of birds.¹² In their obliviousness to the speaker's plight, these marine fowl demonstrate how the world has quite literally become cold to him. Ice and snow engulf a seascape where gulls perform their animal rituals ("baþian brim-fuglas, brædon feþra"), insensible to the friendless exile in their midst. Later in the poem, frigidity deepens into animus as the encroaching dark hurls tempests against humankind: "nipeþ niht-scua, norðan onsendeþ / hreo hægl-fære, hæleðum on andum" (night-shadow darkens, from the north sends bitter hailstorms, with enmity against men, 104–5). Similar moments occur in *Sir Gawain and the Green Knight*. Of the surviving Middle English poems, this is the one most obsessed with interweaving alien nature and human narratives.¹³ This suturing occurs not only in the famous hunt scenes but throughout the unfolding action, in the smallest details: even the Green Knight is decorated with a vegetative flourish, with holly. Every action undertaken by Gawain, every emotion experienced by this knight has some counterpart in the animal-rich landscapes across which the poem unfolds. Gawain's departure from the Arthurian court is the culmination of a kaleidoscopic change of seasons, a sinuous movement from bright spring to desolate winter. A warm sun's advent finds the earth ardent for life, and flora stir with yearning: "When þe donkande dewe dropez of þe leuez / To bide a blysful blusch of þe bryȝt sunne" (When the dew at dawn drops from the leaves, / To get a gracious glance from the golden sun, 519–20). The faded austerity of winter follows, as "Wroþe wynde of þe welkyn wrastelez with þe sunne" (wroth winds in the welkin wrestle with the sun) and the green grass grows grey (525–27) in a universe of change and motion. Gawain, winter cold within his soul, gloomily sets forth from Camelot to discover the habitation of the Green Knight. The nadir of his journey finds him wandering a Welsh wasteland, miserably alone. We know the knight's psychic turmoil not because he voices it like the Wanderer, but because his emotions *are* the landscape. His gloom is an arboreal tangle, his despondency an animal lament:

> By a mountain next morning he makes his way
> Into a forest fastness, fearsome and wild;
> High hills on either hand, with hoar woods below,
> Oaks old and huge by the hundred together,
> The hazel and the hawthorn were all intertwined
> With rough raveled moss, that raggedly hung,
> With many birds unblithe upon bare twigs
> That peeped most piteously for pain of the cold.
>   (740–47)

Gawain, his mind as crowded with foreboding thoughts as an ancient forest, wanders a terrain that at once seems too full (trees by the hundreds) and too bare (winter has stripped everything, invading the whole world with its chill). The knight from Camelot knows that time to accomplish his quest is trickling away, while he remains "mon al hym one" (a man all alone, 749). His misery is embodied in those "mony bryddez vnblyþe . . . þat pitosly piped for pyne of þe colde." In a wordless avian complaint, Gawain's pangs at his solitude find their most lyrical expression.

Animals and Allegory

Among the most influential texts on animals bequeathed to the Middle Ages was the *Physiologus*. Composed originally in Greek, perhaps at cosmopolitan Alexandria, this work circulated widely by the close of the fourth century and eventually inspired the medieval bestiary tradition. A multicultural compendium, the *Physiologus* gathers in its encyclopedic entries science and folklore from Egyptian, Roman, Greek, Jewish, and Indian sources. Early translations of the book survive in Ethiopian, Syrian, Armenian, and Latin, while later versions include Old English, Old High German, Icelandic, Flemish, Russian, and Provençal.[14] The appeal of the *Physiologus* was in part its transmutation of animals and natural phenomena into symbols of biblical truths. Spurred by the impulse to place human signification in animal forms and never content simply to moralize, this text and its medieval progeny discerned in both quotidian and fantastic beasts divine allegories. The owl, for example, figured the Jews, whose refusal to see the light of Christ's truth had doomed them to perpetual night. The Phoenix, reborn from its own ashes, was a type for Jesus, similarly come back from the dead. The aristocratic stag was a Christ-figure vanquishing diabolical snakes; the conniving fox, a deceiving heretic, tracing his errant way; the resplendent panther, his mouth breathing forth divine fragrance, a figure for Christ enjoining his apostles to disseminate his message throughout the world; the industrious bees, obedient, orderly citizens.[15]

Several of the animals described are noted for their sexual habits. The female viper, for example, is said by the *Physiologus* to possess a human form from the face to the navel, and a crocodile's body thereafter. Because she does not possess genitals, the viper must have oral sex with her partner. After drinking the semen of her mate, she castrates and kills him. Because she has no vagina, however, the young vipers engendered through this union must rip their way through her belly, ending her life. The entry concludes by describing the viper as—what else?—a figure for the Jews, who cannot think symbolically, but think only in literal terms. Jews practice circumcision on their flesh, for example, rather than interpreting this ritual of covenant with God as an act to be undertaken spiritually (that is, metaphorically). The viper thereby assists in the work of articulating Christian identity, by distinguishing it from the Judaism from which Christianity historically emerged.

The weasel is similarly said to copulate orally, though she gives birth through her ears (the right ear for boys, the left for girls). Weasels, we are told, are a figure for those who allow wicked sayings to enter their minds and engender sin. The beautiful unicorn cannot be captured by hunters, but should a chaste maiden offer her lap he is happy to lay his head there; lest the image become suggestive, however, we are immediately told that the unicorn is Christ, the virgin is Mary, and there is (by implication) nothing sexual about this strange equine's ardor for placing his long horn in maidenly harbors. The *Physiologus*, like much early Christian writing, stresses the value of chastity and the dangers of desire. Indeed, one of its largest animals is most notable for a complete absence of amorous feeling. The elephant and his wife symbolize Adam and Eve, who never desired each other and possessed no knowledge of coitus before the snake led them into temptation. Elephants mate only out of necessity, and even then would not be able to copulate without the use of an aphrodisiac, namely, mandrake root. Elephants are in this way the purest of animals and an inspiration to continence. Yet it is difficult to avoid wondering whether the eroticism of the viper, the weasel, and the unicorn can be displaced so easily through allegorization. Surely one of the appeals of the *Physiologus* is its narration of fellatio, promiscuity, hermaphroditism, and homosexuality—even if the text ultimately transforms its animal enactors into tidy Christian morals.

Perhaps the most startling of the animals-become-allegories is the hyena (*yena*), yet another figure for the Jew. According to Clement of Alexandria (d. ca. 215) and many later writers, the lewd hyena possesses the sexual organs of both genders and employs them promiscuously.[16] A devourer of corpses and mimic of human language, the hyena with its sexual shiftiness is said to symbolize Jewish duplicity, since the sons of Israel had once worshipped the true God but abandoned him to pursue vices and idols. The hermaphroditic hyena passed from the *Physiologus* into the medieval bestiaries, where it was frequently illustrated. Debra Hassig observes that depicting the Jews as hyenas stresses the "terrestrial Church's ultimate dominion over the 'bestial' Jews."[17] A scene from one bestiary, Bodley 764, represents a hyena feasting upon Christian cadavers in a graveyard guarded by an ornate church. This text, like the *Physiologus* before it, compares Synagoga to an unclean animal. The Aberdeen Bestiary likewise depicts

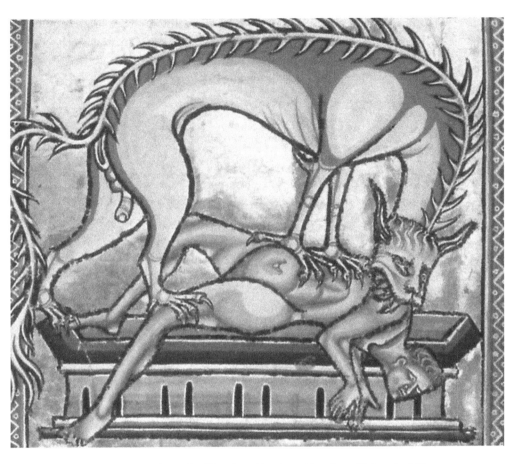

Figure 1. Hyena from the Aberdeen Bestiary. Aberdeen University MS 24, fol. 11v. By kind permission of Aberdeen University.

the hyena visually as well as verbally (fig. 1). This manuscript illustration underscores the animal's utter strangeness by bestowing upon it an enlarged, anally positioned vagina, a penis, an ostentatious spine that whirls into an elaborate tail, and a diabolical head with fearsome fangs. Sinuous crimson markings bestow upon the animal a grim kind of grace. With raptor-like claws the hyena perches upon a fresh corpse, biting the back with gusto. Since the text informs us that the hyena is a symbol for the Jews, their possible sexual alterity may be intimated through the vagina and their association with the devil through the demonic facial features and horns. Most noteworthy, however, are this hyena's oversized, circumcised male genitals.[18] Medieval Christians did not practice circumcision, a bodily modification that was intimately entwined with contemporary imaginings of race.

Animals of Race

To be of a subaltern race has been frequently represented throughout human history as to be of a lesser species. When a group of people lose the privilege of possessing humanity they may be enslaved, dispossessed, or eradicated without moral qualm. Francesca Royster labels this process of reduction as "racing the animal/human divide."[19] We see a transformation of this kind in the Aberdeen Bestiary, where the Jew becomes a hyena. We see a more literal interchange from race to species in Adolf Hitler's pronouncement, "The Jews are undoubtedly a race, but they are not human."[20] Cary Wolfe has coined the useful phrase "discourse of animality" to designate the bestial "constellation of signifiers" that structure "how we address others of whatever sort (not just nonhuman animals)." Though built on the assumption of an ontological divide between the human and the animal, the discourse of animality has, Wolfe notes, "historically served as a crucial strategy in the oppression of *humans* by other humans."[21]

For the Greeks and Romans, civilization demarcated human from barbarian; very little separated the barbarian from the beast. Medieval representation of race as species followed ample classical precedent. The British polemicist Gildas mixed leonine, canine, and lupine traits when he called the Saxons "a race hateful both to God and men" and described their invasion of Britain in the fifth century as "a multitude of whelps [which] came forth from the lair of a barbaric lioness."[22] Because his sympathies lay with the Britons, the Anglo-Saxon conquest is for Gildas the ruinous advent of a "brood" of "wolfish offspring" and "bastard-born comrades," fiercely devouring the land's bounty with "doggish mouths." In composing a sympathetic history of these supposedly subhuman invaders—invaders from whom he was in fact descended—the venerable Bede relied upon Gildas's text but silently dropped his bestializing rhetoric. Geoffrey of Monmouth, on the other hand, was happy to restore the degradation when in the twelfth century he renarrated insular history from a non-Anglocentric point of view.

Most medieval animalizations of human groups were meant to be received and were indeed understood metaphorically. When Geoffrey of Monmouth writes in *History of the Kings of Britain* that the Trojan leader Brutus has surprised the slumbering Greeks with his army, he employs an ancient comparison that enfeebles the enemy while valorizing the heroes: "when [the Greeks] set eye on those who were about to butcher them they were stupefied, like sheep suddenly attacked by wolves."[23] Later in the story the Poitevan leader Goffar will attempt to portray the Trojans as dull herd animals ("we shall seize hold of these weaklings as if they were sheep," 69), but Geoffrey ensures that we see his beloved people only in the fiercely lupine terms that Brutus espouses. Wolves, after all, are associated in his text with agents of natural justice, devouring a sodomite king who exerts his tyranny over the Britons (78). Christians in general can be called sheep ("Christ . . . laid down his life for His sheep," 131), but the Britons are never lamblike until their later, inglorious days on the island, when the Saxons exert greater control over the miserable remnant of this once heroic people.[24]

The lycanthropy Geoffrey worked upon the British was no doubt intended to counteract the contemporary representation of the Welsh as *plus fol que bestes an pasture*, "by nature stupider than beasts in the field."[25] John of Salisbury dismissed the occupants of Wales as "a people rude and untamed . . . they live like beasts and despise the Word of Life."[26] The chronicle known as the *Gesta Stephani* describes the country in fuller but not more sympathetic terms:

> Now Wales is a country of woodland and pasture, immediately bordering on England, stretching far along the coast on one side of it, abounding in deer and fish, milk and herds; but it breeds men of an animal type [*hominum nutrix bestialium*], naturally swift-footed, accustomed to war [*consuetudine bellantium*], volatile always in breaking their word as in changing their abodes.[27]

Here we are dealing with something more than metaphor. If the Welsh truly are no better than beasts, then the English movement into Wales can be justified, since the indigenous population are no more entitled to the land than any other wildlife that happens to move through its untamed landscapes.

Because animals were considered to be reasonless and therefore did not pose troubling questions of agency and autonomy; because God had given primal stewardship of beasts to Adam; and because medieval peoples lived closely with animals and trusted that the proof of human superiority lay in humanity's ability to domesticate, commodify, and profit from the world's fauna, it was perhaps inevitable that animals became receptacles for uneven discourses of race. It would be exaggerating only slightly to say that the representational matrix for medieval concepts of race derived from a discourse of animality rather than a vocabulary for human variation. Animal bodies give to the idea of race a stability that it would not otherwise possess. They arrest the fluctuations of human identity by rigidly embodying it. Animal skin could therefore become a kind of prison. Yet animals might also offer bodies through which could be dreamt identities for which no vocabulary yet existed.

Animal Innovation

> *The animal story's invitation to pleasure is invariably an invitation to a subversive pleasure. . . . It is the very instability of the anthropomorphized animal's identity which can make contact or even proximity with it so hazardous for those with an overblown sense of their own importance, power and identity.*
> —Steve Baker, *Picturing the Beast*

The twelfth-century historian William of Malmesbury was the child of conquest, the biological product of the events initiated by William of Normandy in the preceding century. His mother indigenous and English, his father a French-speaking Norman, and his world a celibate Latinate monastery, William seems to have had a great deal of difficulty working out exactly what history's impurities had fashioned in him. Writing in the turbulent wake of England's transformation from a relatively homogeneous kingdom to a racially bifurcated one, William attempted in his *History of the English Kings* to provide his fractured country with a continuous history. This repair work was in part propelled by his own precarious position as a man caught between worlds that had moved closer together in his lifetime but were far from constituting a unity.[28]

Prized by contemporary scholars as a seemingly modern historiographer, William has at the same time been condemned by critics who do not see the point of the marvels that from time to time erupt within his otherwise sober narrative. Among these wonders is a report of two elderly women who delight in imprisoning men in the bodies of animals and selling them at the market to embark on their new lives, held captive in identities they never dreamt would be their own (*Gesta Regum Anglorum* 2.171). The story of these two witches is a culmination of a narrative arc instigated by a digression about a necromancer-pope who indirectly triggers the Norman Conquest, bringing its themes of racial and cultural admixture into a more bluntly corporeal register. The story of the Circe-like witches emphasizes the confusions engendered when amalgamated identities inhabit a single body, and stresses as well the purifying power of Roman speech:

> On the high road that leads to Rome lived two old crones, altogether filthy and given to liquor, who shared one cottage and were filled with one spirit of witchcraft. If ever a traveller came to lodge with them by himself, they used to make him take the shape of horse or hog or some other animal and offer him for sale to the dealers in such things, spending on their stomachs the coin they thus obtained. One night, as it happened, they gave lodging to a youth who earned his living as an acrobat, and made him to take the shape of an ass, thinking that donkey whose astonishing capers could hold the attentions of passers-by would be a great addition to their assets; for whatever movements were dictated by one of the dames the donkey followed. He had not, you see, lost a man's intelligence, though he had lost the power of speech. (2.171)

Though the women find the performing donkey a lucrative addition to their livestock business, they eventually sell the animal to an avid buyer. The new pet entertains at his owner's drunken feasts, but eventually the novelty of the performing beast diminishes. The crones had warned the donkey's purchaser not to allow the creature to approach water. Now that he is unguarded, the ass runs to a pool and rolls in its cleansing embrace. He is restored immediately to his human shape. Soon thereafter the acrobat is asked if he happened to have seen an escaped donkey wander by. He avers that he was until recently that very animal. The case of the unwanted transformation eventually makes its way to the pope, and the witches are convicted for their crimes.

The acrobat's intelligence (*intelligentia*) at work in a body that renders him strange combines the human and the animal in novel ways, at once entrapping and delighting: he yearns for the contours of an ordinary form while performing feats of invention that neither ass nor man could do alone. The artist in a donkey's skin is a figure of racial and cultural hybridity, of a neoteric and medial state of being which traditional vocabularies of identity were ill equipped to express. Possessed of a fascinating vitality, this conjunction of identities within a single body offers a powerful (if temporary) resolution to all the anxieties about mixed race that circulate in William's narration of his own heritage, anxieties allayed but not transcended through William's ardent embrace of monastic and classical Latin over indigenous English and imported French. True, the man in his animal skin lacks the power of speech (*amiserat loquelam*), and true, he must submerge himself in the purifying power of water to gain the ability to describe his compoundedness, yet William's breathless narration betrays a deep-seated fascination on his part, a lingering enchantment that no restoration to human ordinariness can allay.

As living, moving beings not well inclined to the stasis required by epistemological systems, animals are at best imperfect allegories. They are, in the words of Steve Baker, "the metaphor that cannot hold" (*Picturing the Beast*, 140). This inherent instability affects (indeed, infects) the human as much as the beast. In describing the power of animals to lead us away from the merely human, the late Jacques Derrida punningly spoke of the "animals-word," *l'animot*. A jolting neologism, the French term combines a singular article with an ending that sounds plural but cannot be; it also hybridizes animal (*l'animal*) with meaning-making (*le mot*) as a way of undermining human subordination. "*Ecce animot*," writes Derrida, offering a word forged from the proximate and the "radically foreign, a chimerical word" that—like the classical Chimaera—possesses a "monstrousness derived precisely from the multiplicity of animals" from which it was formed, from the "heterogeneous elements within a single verbal body."[29] William of Malmesbury's narrative of transformation, storytelling, and bodies in flux offers a medieval counterpart to Derrida's *l'animot*. Despite his own hybridity, William was capable of speaking about the past of his beloved homeland with great confidence. Englishness is never examined so much as assumed; it dominates, collects, purifies. Strange figures of impurity and hesitation, like the acrobat-donkey, provide another version of that past; they represent branching and ambiguous paths that if acknowledged could disrupt the chain of history once again, ruining William's careful repair work. William does not follow these uncertain roads to their unknown destinations, but prefers stable histories that lead to secure futures. He distances his wondrous bodies geographically or through their gender. Despite allowing contradiction in his narrative, and despite his acknowledgment that history is messy, incongruous, and difficult to sort out, the *History of the English Kings* ultimately sides with continuity and firm foundation over invention and disruptive innovation. The acrobat-donkey is abandoned at the wayside along the road to Rome.

The Animal in the Middle

Animals in William of Malmesbury are energetic figures of that which does not yet possess a precise lexicon. They are medial bodies, denizens of a restless space where a troubled past meets an uncertain present, headed toward but never quite arriving at their unknown, inarticulate future. Yet William dismisses or abandons each *l'animot*, each attempt to articulate identities that he—like the acrobat inside the donkey—lacks the words to describe. In these bodies that conjoin the human to the animal, however, can be glimpsed possibilities not yet delimited, possibilities that endure long after their narratives end. Alphonso Lingis has written eloquently of this animal of possibility that we carry within, and the ways in which this inner alterity scatters the human and renders it porous, a swarming and difference-ridden multiplicity:

Our bodies are coral reefs teeming with polyps, sponges, gorgonians, and free-swimming macrophages continually stirred by monsoon climates of moist air, blood, and biles. Movements do not get launched by an agent against masses of inertia; we move in an environment of air currents, rustling trees, and animate bodies. . . . Our legs plod with elephantine torpor; decked out fashionably, we catwalk; our hands swing with penguin vivacity; our fingers drum with nuthatch insistence; our eyes glide with the wind rustling the flowering prairie.[30]

Donna Haraway, a former theorist of cyborgs and current theorist of interspecies comminglings, similarly writes that dogs—like any animal that has become our queer companion—"are not about oneself. . . . They are not projection, nor the realization of an intention, nor the telos of anything" (*Companion Species Manifesto,* 11). The multiplicitous interrelationships in which animals and humans find themselves entangled amount to "ontological choreography, which is that vital sort of play that the participants invent out of the histories of the body and mind they inherit and that they rework into fleshly verbs that make them who they are. They invented this game; this game remodels them. . . . The word is made flesh in mortal naturecultures" (100). The animal fact infects the real and breaches the boundary that was supposed to separate both, a phenomenon that might be christened *aninormality.*

This animal infection, this alliance-based escape from the contours of a merely contextual reading, can likewise be glimpsed in the Aberdeen Bestiary, in the corpse-eating beast that on a first reading is clearly an anti-Semitic dismissal of a group of people who probably lived among the book's makers. Yet from another point of view the creature is so strange that it seems to have stepped off the pages of surrealist automatic writing, or perhaps out of the frame of some painting by Dali. This animal, this monster, is what in Latin is called simply the *yena,* but in English could be christened the bisexual Jewish hyena. Aberdeen University MS 24 is a beautiful manuscript of unknown authorship and patronage. Insular and probably produced late in the twelfth century, the bestiary yields few clues to its historical context, but Lincoln and York have been proposed as likely sites for its manufacture.[31] Both these cathedral cities harbored a significant Jewish population; both witnessed violence by Christian residents against these Jews. The text can and should be located within a time and a place grown perilous for the people whom the hyena figures. In 1190 Lincoln witnessed hostility against its Jewish population: houses were pillaged and their residents were forced to seek harbor in the castle. Not long thereafter York was the site of the infamous ha-Gadol mass suicide and murders. This bloody event unfolded after the Jewish families of the city had been forced to take refuge at Clifford's Tower. The Jews who did not take their own lives there were massacred as they attempted to leave.[32] When the bestiary declares that *filii Israel* (the sons of Israel) resemble the hyena, with its rigid spine and deceitful ways, we would do wrong to see in this equation some disembodied or innocuous representation. By repeating an ancient slander during troubled times, the text and its image carry hate and provoke violence. Representing Jews as a hyena has potentially bloody repercussions, and may even be complicit in contemporary pogrom.

Yet the scene as narrated and especially as illustrated overpowers its historical context. The Jews do not vanish but through their unlooked-for alliance with the hyena become a way of seeing beyond and outside the suffocation of being merely Christian, of being merely human. With its body that refuses to obey the binary gender system, with its mouth full of Christian flesh being devoured and incorporated into a monstrous form that was supposed to be the unthinkable limit of *Christianitas,* the *yena* becomes a version of Derrida's *l'animot,* Lingis's oceanic humanity, and Haraway's companion species: an invitation to explore a spacious corporeality beyond the specious boundaries of the human, to invent through alliances with possible bodies a monstrous kind of becoming that carries history within but which is not reducible to historical allegory. The bisexual Jewish hyena cannot in the end be explained away by a singular cultural use, or reduced to a local story, or dead-ended by that hatred which attended its birth and which may have been reinvigorated by its twelfth-century English release. Limned by violence but not fully enlimed by it, the hyena of the Aberdeen Bestiary emplaces the animal at that cognitive edge where the drag of mere context, of historical determination, must yield to the pull of dreamier horizons and unforeclosed possibilities. The hyena's story is surely one both of human limitation and of the possibility of seeing beyond such limitation to a space where all that seemed natural, pregiven, and unchangeable begins to lose its solidity and its savagely delimiting force.

Even if the animal thereby seems to reach a space where history can no longer determine the totality of its story, is it fair to say that such a tenuous space is any less anthropocentric than the allegories and contextual readings that would keep a beast prisoner to the bestiary? *Sir Gawain and the Green Knight* featured a protagonist's movement across geographies as kinetic and as animated as the Aberdeen hyena's trajectories. In the somber errantry of Gawain we find not only an instance of the pathetic fallacy, where anthropocentrism leads a human author to glean human meanings from a nonhuman landscape, but also what the philosopher Gail Weiss has called "embodiment as intercorporeality." We can witness the ways in which our identities are dispersed across the relations we form.[33] Even better, this vegetal and animal dispersedness could be termed an *interspecies alliance*, the mode by which a knight of the Arthurian court can share his sorrow at the world's chill with birds huddled in winter misery. These animals give voice to their sadness in a language that, while not human, is also not so very difficult to understand:

> þe hasel and þe haʒþorne were harled al samen
> With roʒe raged mosse rayled aywhere,
> With mony bryddez vnblyþe vpon bare twyges
> þat pitosly þer piped for pyne of þe colde.
> þe gome vpon Gryngolet glydez hem vnder,
> þurʒ mony misy and myre, mon al hym one.
>
> (744–49)

> [The hazel and the hawthorn were all intertwined
> With rough raveled moss, that raggedly hung,
> With many birds unblithe upon bare twigs
> That peeped most piteously for pain of the cold.
> The good knight on Gringolet glides thereunder
> Through many a marsh, a man all alone.]

We know already that this last line must be untrue. Despite what at first glance appears to be his somber solitariness, Gawain's subjectivity is entangled in hazel and hawthorn, his embodiment completed by a flock of shivering birds, his knightly identity inseparable from his good steed Gringolet. The romance creates a space where embodiment is multiple and interspecies. Sir Gawain glides through a world alive with flora and fauna, all with their own agency, a world where the knight can never be *mon al hym one.*

The medieval animal is, like the human animal, a creature of history. Culture and context conspire to delimit its meanings and relegate it to a state of being that is subordinate, silent, still. Though at times an excellent vehicle for the expression of human meanings and stories, though at times a body that seems too easily reduced to context and anthropomorphic determination, the animal—no matter how intimate it becomes to human worlds—remains apart, persists as strange. The animal that (as Derrida would say) therefore I follow leads us to a middle space where allegories and moralizations seem insufficient in their power to contain. It is a place of dispersal, multiple agency, and intercorporeality. Leaving the human less confident of integrity and dominion, the animal goes its own way, inviting us to see what happens when the species line fades, when the inhuman inside us invites us toward unknown horizons: the liminal space of the human and the posthuman, of the ancient and the utterly new—the threshold called the medieval.

Notes

My gratitude goes to Susan Crane and Karl Steel of Columbia University and Gail Kern Paster of the Folger Shakespeare Library for catalytic conversations on medieval animals. This essay owes much to its first audience at Ohio State University (especially Barbara Hanawalt, Ethan Knapp, and Lisa J. Kiser) and to those at Princeton and Bucknell who helped its argument to burgeon. Finally, thanks are due Michael Craig and the University of Aberdeen.

1. I have specifically in mind here Cary Wolfe's edited collection *Zoontologies: The Question of the Animal* (Minneapolis: University of Minnesota Press, 2003) and his monograph *Animal Rites: American Culture, the Discourse of Species, and Posthumanist Theory* (Chicago: University of Chicago Press, 2003); Jennifer Wolch and Jody Emel, eds., *Animal Geographies: Place, Politics, and Identity*

in the Nature-Culture Borderlands (London: Verso, 1998); and the second edition of Steve Baker's seminal *Picturing the Beast: Animals, Identity, and Representation* (Champaign: University of Illinois Press, 2001). Donna Haraway has traded in cyborgs for interminglings of dog and human in her wonderful *Companion Species Manifesto: Dogs, People, and Significant Otherness* (Chicago: Prickly Paradigm Press, 2003). A theorist whose writings on animals and nature now seems prescient is the surrealist biologist Roger Caillois; see my post on him at http://jjcohen.blogspot.com/2006/02/roger-caillois-and-medieval-animals.html (much additional writing on medieval animals may be found on this blog, especially the rich posts by Karl Steel). Although I disagree with her central thesis, the materials Joyce E. Salisbury collects in *The Beast Within: Animals in the Middle Ages* (New York: Routledge, 1994) are excellent, as is her analysis of animals and sexuality. Also of great use to medievalists are the collection edited by Angela N. H. Creager and William Chester Jordan, *The Animal/Human Boundary: Historical Perspectives* (Rochester: University of Rochester Press, 2002), and David Salter's *Holy and Noble Beasts: Encounters with Animals in Medieval Literature* (Cambridge: D. S. Brewer, 2001).

2. On "possible bodies" see my *Medieval Identity Machines* (Minneapolis: University of Minnesota Press, 2003), xi–xxiii. Throughout this essay I will be using "Middle Ages" rather loosely, but (as will become clear) I mean mainly the European, Latin Middle Ages, with a special emphasis on twelfth-century Britain. Much of what I say is general enough not to hold true in culturally specific cases; see, for example, the nuanced discussion of how nature is different in Welsh and Irish sources contrasted with Anglo-Saxon ones, in Alfred K. Siewers, "Landscapes of Conversion: Guthlac's Mound and Grendel's Mere as Expressions of Anglo-Saxon Nation-Building," *Viator* 34 (2003): 1–39 (animals are specifically discussed at 19). A book that usefully makes similar points about the changing construction of nature especially in its relation to empire is Richard H. Grove, *Green Imperialism: Colonial Expansion, Tropical Island Edens and the Origin of Environmentalism, 1600–1800* (Cambridge: Cambridge University Press, 1995).

3. Geoffrey Chaucer, *The Riverside Chaucer*, ed. Larry D. Benson, 3rd ed. (Boston: Houghton Mifflin Company, 1987), I.1308. Further references to the Canterbury Tales are by fragment and line number in this edition.

4. Robert Bartlett writes that naming a pet with a nonhuman appellation "simultaneously brings humans and (some) animals closer and distances them": *England under the Norman and Angevin Kings, 1075–1225* (Oxford: Clarendon Press, 2000), 668. He offers as some examples of pet names Lym and Libekar for greyhounds; Liardus for a horse; and Gibbun, Refuse, and Blakeman for falcons. For Chaucer's dogs, see the Nun's Priest's Tale 3383, where Colle is referred to in a domestic plural ["Colle oure dogge"] that evidences

the affection felt for him. Chaucer supplies a third canine appellation that may be derived from a family name, Talbot, probably a "kind of hunting dog introduced into England by the Talbot family" (*Riverside Chaucer*, p. 941n).

5. Midas Dekkers has an intriguing chapter ("Vital Juices") about this transgression of species boundary through such fluid intermingling: *Dearest Pet: On Bestiality*, trans. Paul Vincent (London: Verso, 1994). In an unforgettable turn of phrase Dekkers describes the milk of other mammals as "intimate udder juice . . . a drinkable form of Esperanto" (94).

6. *Gesta Regum Anglorum: The History of the English Kings*, 2 vols, ed. and trans. R.A.B. Mynors, completed by R.M. Thomson and M. Winterbottom (Oxford: Clarendon Press, 1998), 5.409.

7. On William's and Orderic's condemnations, see David Crouch, *The Normans: The History of a Dynasty* (London: Hambledon and London, 2002), 168. Crouch initially treats Woodstock at 159 (describing a young Henry's feeling of being marginalized at court, the sympathy he perhaps felt for the denigrated English, and the satisfaction he may have found in the surrogate household he created at the royal hall). He later describes Woodstock's transformation from an "alternative [i.e. English] household" into a menagerie, probably because the English concubines and children previously kept there had moved along or grown up (193).

8. The convergence of the classical and biblical traditions of human superiority is traced in Bartlett, *England under the Norman and Angevin Kings*, 681.

9. On dominance, power, and human differentiation from animals, see Karl Steel's excellent essay "'Elles were beest lich to man': The Problems and Uses of Animal Likeness in *Sidrak and Bokkus*" (*Exemplaria*, forthcoming).

10. From Thomas of Chobham's manual of instruction on preaching, *Summa de arte praedicandi*, chapter 7; trans. D.L. d'Avray in *The Preaching of the Friars: Sermons Diffused from Paris before 1300* (Oxford: Clarendon Press, 1985), 232–33. Debra Hassig quotes the passage as her epigraph to *Medieval Bestiaries: Text, Image, Ideology* (Cambridge: Cambridge University Press, 1995), xv, finding it an "eloquent" expression of "the medieval belief that the natural world of beasts and birds is a book of lessons written by God for the edification of human beings."

11. "Omnis mundi creatura / quasi liber et pictura / nobis est et speculum." Alan of Lille, *De Incarnatione Christi*, PL 210, 579A; cited by Nona C. Flores in *Animals in the Middle Ages* (New York: Garland, 2000), ix.

12. *The Wanderer*, ed. R.F. Leslie (Manchester: Manchester University Press, 1966).

13. *Sir Gawain and the Green Knight*, 2nd ed., ed. Norman Davis (Oxford: Oxford University Press, 1967). Translation from Marie Borroff, *Sir Gawain and the Green Knight* (New York: Norton, 1967).

14. Michael J. Curley surveys the provenance and dissemination of the text in the introduction to his translation of the Latin *Physiologus* (Austin: University of Texas Press, 1979), xvi–xxix. For the Latin version see Francis Carmody, ed., "Physiologus Latinus, versio Y," *University of California Publications in Classical Philology* 12 (1941): 95–134.

15. Debra Hassig explores these symbolic meanings and many others from the bestiary tradition in *Medieval Bestiaries*, stressing that although some allegorical readings were fairly widespread, almost all animal bodies were open to multiple interpretations, often within the same text.

16. John Boswell translates the passage from Clement and provides a useful commentary in *Christianity, Social Tolerance, and Homosexuality: Gay People in Western Europe from the Beginning of the Christian Era to the Fourteenth Century* (Chicago: University of Chicago Press, 1980), 356.

17. Hassig, *Medieval Bestiaries*, 150.

18. For this example and for an excellent overview of Jews as "blind beasts," see Anthony Bale, "Fictions of Judaism in England before 1290," in *The Jews in Medieval Britain: Historical, Literary and Archaeological Perspectives*, ed. Patricia Skinner (Woodbridge: Boydell Press, 2003), 141–42. Bale writes perceptively of the bestiary tradition that it "offers us clear evidence that Jews' bodies (as well as the Jewish religion) were thought of as degraded and corrupt entities, foreshadowing later antisemitic material and a 'racial' conception of Judaism" (141). In *Medieval Bestiaries* Hassig connects the prominent genitals of the Aberdeen hyena to contemporary worries about Jewish-Christian intercourse, often figured as bestiality. See also Bettina Bildhauer and Robert Mills, "Introduction: Conceptualizing the Monstrous," in *The Monstrous Middle Ages*, ed. Bildhauer and Mills (Toronto: University of Toronto Press, 2003), 12. Steven Kruger writes of the lingering tinge of animality thought to adhere to both Saracen and Jewish converts to Christianity, as part of his excellent treatment of race in "Conversion and Medieval Sexual, Religious, and Racial Categories," in *Constructing Medieval Sexuality*, ed. Karma Lochrie, Peggy McCracken, and James A. Schultz (Minneapolis: University of Minnesota Press, 1997), 158–79.

19. Francesca Royster, "'Working Like a Dog': African Labor and Racing the Human-Animal Divide in Early Modern England," in *Writing Race Across the Atlantic World: Medieval to Modern*, ed. Philip D. Beidler and Gary Taylor (New York: Palgrave Macmillan, 2005), 113–34. See also Charlie Leduff's article from the *New York Times*, "At a Slaughterhouse, Some Things Never Die," reprinted in the anthology *Zoontologies*, ed. Cary Wolfe, 183–97. I have written at greater length about medieval race and its relation to animality in *Hybridity, Identity and Monstrosity in Medieval Britain: On Difficult Middles* (New York: Palgrave Macmillan, 2006), especially 11–42.

20. Taken as the epigraph to his book *Maus: A Survivor's Tale* by Art Spiegelman (London: André Deutsch, 1987) and treated by Steve Baker in *Picturing the Beast*, 141.

21. "Introduction," in *Zoontologies*, xx. Wolfe also cites Étienne Balibar's "Racism and Nationalism." Balibar writes that racism is undergirded by "the persistent presence of the same 'question': that of *the difference between humanity and animality* . . . the systematic 'bestialization' of individuals and racialized human groups" (Étienne Balibar and Immanuel Wallerstein, *Race, Nation, Class: Ambiguous Identities*, translation of Balibar by Chris Turner [London: Verso, 1991], 56).

22. Gildas, *De Excidio Britonum*, ed. and trans. Michael Winterbottom (London and Chichester: Phillimore, 1978), 23.

23. I quote from the translation of Lewis Thorpe (London: Penguin Books, 1966), here at 61. For the Latin see *The Historia Regum Britannie*, vol. 1, *Bern, Bürgerbibliothek MS 568 (the 'Vulgate' Version)*, ed. Neil Wright (Cambridge: D.S. Brewer, 1984).

24. In fact the "sheepification" of the Britons occurs twice, once after the nobility has relocated to Brittany and left behind men who act like "sheep wandering about without a shepherd" (146; cf. 166), and then after the glory days of Arthur have ended and plague has decimated the island ("'Thou hast given us, O God,' they shouted, 'like sheep for meat,'" [281]).

25. Chrétien de Troyes, *The Story of the Grail (Li Contes del Graal or Perceval)*, ed. Rupert Pickens, trans. William W. Kibler (New York: Garland, 1990), 242.

26. *Letters of John of Salisbury*, vol. 1, *The Early Letters (1153–1161)*, ed. W.J. Millor and H.E. Butler, rev. C.N.L. Brooke (London: Thomas Nelson and Sons, 1955), 135.

27. *Gesta Stephani*, ed. and trans. K.R. Potter, introduction and notes by R.H.C. Davis (Oxford: Oxford University Press, 1976), 1.8.

28. I treat William of Malmesbury's self-conflicted identity at greater length in *Hybridity, Identity and Monstrosity in Medieval Britain*, 54–63. See also the penetrating analysis of William by Laurie A. Finke and Martin B. Shichtman in *King Arthur and the Myth of History* (Gainesville: University Press of Florida, 2004).

29. Jacques Derrida, "The Animal That Therefore I Am (More to Follow)," trans. David Wills, *Critical Inquiry* 28 (2002): 369–418; quotations at 405, 409, 415. See also Derrida's essay "And Say the Animal Responded?" trans. David Wills, in *Zoontologies*, 121–46, as well as Steve Baker, "Sloughing the Human," in *Zoontologies*, 147–64.

30. Alphonso Lingis, "Animal Body, Inhuman Face," in *Zoontologies*, 165–82, quotation at 167.

31. Michael Arnott and Iain Beavan review the scholarship in their introduction to the electronic version of the text. See "History" at http://www.abdn.ac.uk/bestiary/ as well as Xenia Muratova, "The Bestiaries: An Aspect of Medieval Patronage," in *Art and Patronage in the English Romanesque*, ed. Sarah Macready and F. H. Thompson (London: Society of Antiquaries, 1986), 118–44.

32. The best account of the events at York remains R. B. Dobson, *The Jews of Medieval York and the Massacre of March 1190*, University of York Borthwick Papers 45 (York: University of York, 1974, rev. 1996).

33. Gale Weiss, *Body Images: Embodiment as Intercorporeality* (New York: Routledge, 1998).

CHAPTER THREE

Ritual Aspects of the Hunt à *Force*

Susan Crane

From the thirteenth into the fifteenth century, hunting treatises in English and French describe a kind of hunt so elaborated and formalized that scholars invoke the term "ritual" to account for it. This essay explores "ritual" as a descriptor for the kind of hunt the treatises value most, *à force* (*with strength*), concentrating on treatises by three men known to have been expert in hunting: William Twiti, huntsman of Edward II (writing or dictating ca. 1325); Gaston III, Count of Foix (writing or dictating 1387–89); and Edward of Norwich, second Duke of York (translating and modifying Gaston's treatise, ca. 1410).[1] I emphasize works by known practitioners in the conviction that their treatises correlate in some measure with their practice in the field, although their practical knowledge does not guarantee that their texts depict "what happens" at every turn. The more accurate assumption, given their mixed goals of instructing, reporting, and moralizing, is that the treatises reveal how these aristocratic hunters conceived the ideal hunt, and what they believed its value and significance to be.[2] The ritual potential resides at this intersection of beliefs and performance, when gestures are understood to be heavy with meaning, and actions are seen as codified repetitions as well as responses to a present circumstance.

"Ritual" might seem an unlikely analytical framework for an unpredictable chase across miles of country. The hunt *à force* looks more like a "cultural performance," the category Clifford Geertz so influentially identified in the Balinese cockfight: a performance that encapsulates and models the aristocracy's most profound convictions about itself and its place in creation. Alternatively, this kind of hunting could be discussed as a "game" that combines most of that term's meanings by 1350—an amusement, a pastime with rules, a field sport, and a physically challenging contest. Cultural performance, game, and ritual overlap and co-inform hunting; no single category perfectly fits and accounts for it. Analogously, on a much larger scale, John MacAloon analyzes the modern Olympic Games as a nested set of performance types: games (the athletic events) that are incorporated into a ritual (a formal articulation of human unity) that is, in turn, staged as a festival (a recurrent celebration designed for pleasure) that is, at the same time, a spectacle (designed to be watched by a public that is not performing).[3] I concentrate on ritual in this essay because it is the least intuitively plausible performance category for hunting, and because I believe the ritual features of this kind of hunting shape it into a powerful assertion of aristocratic superiority.

The first three sections of this essay argue that the hunt *à force* uses the strategies of secular ritual to affirm the rightness of a social and natural hierarchy headed by the aristocracy. The last two sections argue that the hunt's affirmation of hierarchy involves certain intimacies between humans and animals that appear anti-hierarchical. Does contact with animals undermine the aristocracy's performance of supremacy?

Defining Hunt and Ritual

The hunt *à force*, as described in the treatises, is a highly organized pursuit of a large beast that runs well before hounds, is difficult to capture and kill, has positive symbolic associations, and provides meat that is considered edible. On all four measures, the hart (the red deer stag) is the most favored: he flees a long time before exhaustion; he doubles back, covers his scent in water, and uses other elusive maneuvers; and he stands at bay and fights bravely at the end: Gaston's treatise praises the hart in the proverb "after the boar, the physician; after the hart, the bier."[4] Boar are also prized targets for hunting on account of their exceptional ferocity as well as their meat and the long chase they provide. Both hart and boar have strong symbolic value within chivalry, as shown in their frequent appearance on coats of arms and crests, although here too the hart has the edge, through his association with Saints Hubert and Eustace and his putative power over poisonous snakes. The boar is fierce to a fault: Edward refers to his "despitous dedis"; Gaston calls him haughty and proud.[5]

English and French hunting treatises agree on the general shape of the hunt *à force*. Details vary, but do so within a shared conviction that the hunt is a formalized activity for a large following of experts, under the direction of a king or a lesser lord whose hunting park or forest rights enclave land for the specific purpose of hunting. In the hunt *à force*, tracking hounds first locate potential prey, the most favored being the hart in summer and boar in winter. A single, most challenging beast is chosen as the day's prey and pursued on foot and on horseback with the aid of the hounds. A large repertoire of hunting cries and horn calls communicate to both the hounds and the variously positioned hunting party. For example, Twiti instructs that "after the hart is started, you must blow two motes [horn blasts]. And if your hounds do not come back to you as quickly as you would like, you must blow four motes to hasten the hunting party toward you and to inform them that the hart has been started."[6] Chased to exhaustion through all its ruses, the hunted animal is dispatched, and meticulously dismembered in a fashion known to literary scholars from the Tristan romances and *Sir Gawain and the Green Knight*. A reinforcing meal for the hounds and a procession home, sometimes with an ordering specific to the animal hunted, close the event.[7]

The procedures of the hunt *à force* are not chosen for their efficiency at capturing game. From the purely practical point of view, baited traps, snares, and nets are more effective than stalking and chasing a single animal for as long as it keeps running. Indeed, some treatises include instructions on hunting in more efficient ways, such as the bow-and-stable harvesting of deer; Gaston recommends traps and snares to the old, the infirm, and priests; and *Modus et Ratio* recommends them to the poor.[8] In the context of these other methods of killing game, the hunt *à force* stands out the more clearly as an ideologically charged event. Practicality is not an issue, any more than it was for the differently structured deer hunt of my Midwestern childhood, on which a few male companions or a father and sons tracked game together, prizing most the proximity they achieved in their quiet pursuit on foot, until the animal was brought down with gunshot or bow and arrow at close range. Historically charged motives informed this kind of hunt, notably the celebration of American Indian methods and the mimesis of self-reliant masculinity.[9] In short, there are many ways to hunt that result in killing game; like the hunts of my childhood, medieval aristocratic hunts carried an ideological message.

An ideologically informed performance is just one characteristic of ritual. In its classic definition, a ritual is a ceremonial occasion that calls a community together to mark some kind of change. In the process, a ritual invokes sacred forces, or values held sacred by the community; and it recruits all present as active participants. Ceremonial forms frame the ritual temporally and spatially, setting it apart from everyday life and giving it rules or ways of proceeding that make it repeatable and make its unfolding predictable. The familiar illustration is a wedding, a good illustration because weddings show how variously these requirements of ritual can be met. In the marked space of a judge's chamber, a decorated parlor, or a sanctuary, at a designated and published moment in time, two persons change from being single to married. The persistent tactic of weddings, and rituals in general, is mimesis: the couple join hands, exchange rings, have their hands bound together, walk together in a set pattern, and/or exchange a kiss, to represent mimetically their joining together in marriage. The ritual may invoke the power of a deity or the power of values held sacred, such as love and troth. As Moore and Myerhoff observe in their study of secular rituals, "one can imagine a non-religious society, but it is difficult to imagine a society that holds nothing sacred."[10] Whether the wedding ceremony invokes a god or an ideal of undying love to sanction the marriage, participants including the bride and groom have been known to weep under the symbolic weight of the moment.[11] Further, as is characteristic of ritual, participants at weddings are not just spectators; they are recruited to endorse and certify the marriage. Recruitment occurs through the several roles of the wedding party, and at the moment so perceptively stressed in *Jane Eyre* when all are asked if they know of any just cause why the bride and groom may not lawfully be joined together.

More pervasively, as Eve Sedgwick argues, every wedding guest undergoes interpellation in witnessing a wedding vow that can take place only in heterosexual terms (in most of the United States): the "subject gets constituted in marriage through a confident appeal to state authority, through the calm interpellation of others present as 'witnesses,' and through the logic of the (heterosexual) supplement whereby individual subjective agency is guaranteed by the welding into a cross-gender dyad."[12] It is important to keep ritual's ideological coerciveness in mind as the final effect of its more limited practical goals, such as marrying, healing, or consecrating, and its particular tactics, such as framing, formalization, and repeatability.

Is the Hunt à *Force* a Ritual?

The term "ritual" shows up often in scholarly discussions of medieval hunting. But the focus of these studies is not on ritual per se, and the term typically receives little elaboration. Eve Salisbury calls the hunt *à force* "highly ritualized," and Dorothy Yamamoto refers to Edmund Leach's argument that game animals get "ritual value" from the rules that control when and how they may be hunted.[13] John Cummins goes furthest with the concept of ritual, associating it with formalization: "there was a recognized way of doing everything: formulaic cries, commands, and horn-calls; ritualized ceremonies. The most striking imposition of ceremonial on activities essentially practical came after the death, in the flaying and butchering . . . and in the *curée*, the formal rewarding of the hounds."[14]

Rules of behavior and spoken formulas contribute to the stylization expected of rituals, and there are further expectations for ritual that the hunt *à force* also meets. It is a framed activity, set apart from the rest of daily life, in parks and preserves and by forest law. Hunting has a role for all the participants of various ranks; the chief huntsman shapes the event and the highest ranking lord, ostensibly at least, directs it. All these specifications contribute to the repeatability of hunting: it can proceed in many instances along the same lines as it has before. Framing, active participation by a whole group, and repeatability are characteristic of ritual, with the last of these holding a preeminent place. But hunting has elements of unpredictability as well, due to its accretion around a practical effort to kill a wild animal. This practical accomplishment is less susceptible to formalization than marrying a couple or declaring a peace or crowning a king, because the animal cannot be as fully recruited to the rules as humans can be. In compensation, the stages of hunting before and after the chase proper are made the more predictable and repeatable: the specifications for choosing which animal to hunt and for cutting up the quarry and rewarding hounds and hunters are salient, but organizing the return home as a procession is also typical.

Moving from hunting's structure to its purposes, we must recall that typically, rituals make and mark a transformation: undergraduates become graduates, girls become women, enemies become allies, the sick become well. It is not clear that hunting *à force* has this purpose, despite revolving around an animal's death: its relation to sacrifice would be intriguing to consider. However, religious and secular rituals do not always involve transformation. Some are more intent on affirming and celebrating the identity of a group and its core values: affirming and practicing faith, for example, in religious services such as matins and evensong. An example from scouting and children's camps in the United States is the raising and lowering of the national flag with bugle calls, a pledge of allegiance, and special ways of moving and folding the flag. Similarly, several recent presidential conventions of Democrats and Republicans have worked to affirm collective values; the parties' nominees have in fact already been chosen in the primaries, but this has not obviated the importance of the nominating convention. To be sure, it does mark a change—the start of the formal campaign period—but primarily the convention generates a message about the identity of a group, and thereby it attempts to change the group's future prospects. As with weddings, the dominant tactic is mimesis: the nominee must be seen to be "presidential"; he is staged with all others deferring to him, praising him, and reiterating that his victory is assured.

My contention is that the hunt *à force* is a mimetic ritual designed to celebrate and perpetuate aristocratic authority. This kind of hunting defines a miniaturized cosmos within which aristocratic ability, superiority, and governance are represented. In other words, this kind of hunting surpasses merely being another real-life *instance* of skill, superiority, and governance. Instead, the ritualization of hunting *endorses and validates* aristocratic skill, superiority, and governance. It sets up a performance space in which aristocracy mimes its own myth of itself.

Hunting as a Mimesis of Aristocratic Power

The boar hunt *à force* provides a good illustration for this thesis. Plausible and to some extent operant purposes of this hunt are to kill boar and come home with boar meat. These goals could be more easily accomplished by building a strong pen with a one-way gate and placing a nice pile of apples inside. Shooting or spearing the penned boar would be as easy as shooting fish in a barrel. Gaston describes such methods, terming them base and ugly, and adding that "I speak of this against my will, for I should only teach how to take beasts nobly and gently."[15] In ritual hunting, it is the process of hunting, not the size of the bag, that counts. Taking boar nobly involves a long chase with a pack of hounds, horn calls, hunter-retainers, and mounted gentry. Figure 1 illustrates the boar hunt from an early manuscript of the *Livre de chasse*: "He will run well from sunrise to sunset," says Gaston; then he turns and defends himself fiercely, wounding and killing hounds, horses, and men.[16] The ideological image generated by this kind of hunting is no less powerful for its explicitness: the aristocratic household, under the lord's rule, can dominate a fearsome and adversarial world of nature. The hunt's danger makes it appropriate for aristocratic leadership. "It is great mastery and a fine thing to know how to kill a boar with a sword," Gaston exclaims. He particularly admires a boar kill executed from horseback, although he cautions that it is a dangerous method.[17] The value he assigns this method recalls the commonplace that hunting prepares knights for war by testing their courage and horsemanship.[18] The requirement that ritual hunting test courage inspires the instruction that the boar at bay should be urged to charge, with shouts of "Avant, mestre! Avant!" (Come on, sir! Forward!)[19]

The hunt's conspicuous consumption of resources further mirrors aristocratic life more generally—its large households, its lavish accoutrements, and the grand scale of its entertainments. In order to produce a mimesis of social hierarchy and aristocratic authority even over nature, hunting must be a complexly staffed operation that unfolds in

Figure 1. Gaston Phébus, *Livre de chasse*. Paris, Bibliothèque nationale de France, MS français 616, fol. 73. By permission of the Bibliothèque nationale de France.

an orderly way. Riding to hunt is appropriate for the gentle participants, and this may be why beasts that run well are favored targets, even the comparatively scrawny fox. On the significance of riding, Jordanus Ruffus (a knight in charge of Emperor Frederick II's stable) writes in his treatise *De Medicina Equorum* that "no animal is more noble than the horse, since it is by horses that princes, magnates and knights are separated from lesser people and because a lord cannot fittingly be seen among private citizens except through the mediation of a horse."[20] Figure 1, like all of this manuscript's depictions of the hunt *à force*, differentiates the hunting party into a more powerful group on horseback and a lesser group on foot with the hounds.

All these tactics prepare for Sedgwick's ritual interpellation: every member of the hunt not only *has* a role in its performance of social difference, but also is *confirmed* in that role and must *accede* to that role or violate the rules of the chase. Thus the hunt *à force* both expresses and expands the honor of its aristocratic participants: the *Chace dou Cerf* begins, "this sport is so sovereign that any king, or count, or even Gawain, if he lived and loved it well, would gain honor for it."[21] The expansion of honor happens both through the hunt's occasions for courage and through the hierarchical arrangement of roles in the hunting party. James Howe expands on the latter point in his study of modern hunting: "In reference to the ritual of fox hunting, I would argue that it is less a manifestation of a prior consensus than a means by which people are persuaded to consent to the social hierarchy, or if they cannot be entirely persuaded, a means by which they can be induced to give conventional signs of assent."[22] In the boar hunt illustration (fig. 1), those on foot are endorsing their subordination by acting it out with every step—whether or not they would endorse it in principle.

And finally, this mimesis of aristocratic supremacy celebrates nature's as well as society's subordination. The hunt *à force* is designed to last from sunup to sundown. Matching the hunt to the sun's course "imitates the rhythmic imperatives of the biological and physical universe," to quote Moore and Myerhoff again, "thus suggesting a link with the perpetual processes of the cosmos."[23] The treatises further evoke cosmic design by marking seasons for hunting, and especially by presenting the hunt as human domination over wild beasts, an instance of the hierarchical cosmos described in Genesis. It is entirely in keeping with God's plan that humans should rule over and make use of animals. Humans are created in God's image, endowed with reason and speech: as summarized in *Modus et Ratio*, God "created two kinds of animals, ones that he called human animals, and the others were called dumb animals, and they were called dumb because they had no knowledge of the Creator."[24] The formalities of the hunt *à force* construct a microcosmic model of creation, in which creation's hierarchy of humans over animals reinforces the human social hierarchy.

But a salient aspect of the hunt *à force* complicates its performance of aristocratic power: it brings humans into contact with animals. Once a complete earthly hierarchy is posited, with hunters dominating nature, intimacy with animals is potentially abasing. Two kinds of contact appear risky in this regard: talking to hounds and making physical contact with the hunted animal.

The Human-Hound Partnership

Hunting treatises recognize that, although hounds are much better at tracking game than humans, they can learn to hunt on behalf of humans and in the ways humans designate for them. What a triumph of training that a deer hound can learn not to chase just any red deer, not the one that might be straight ahead and a sure thing, but only the great hart chosen at the beginning of the day, who must be pursued through all his evasions even at the risk of getting no quarry at all! Cries and whips together recall this lesson. Edward instructs that if the hounds change from pursuing the chosen hart to a lesser target, "thei shuld be fallen afore and asc[r]ied and wele lasshed" (they should be got in front of and scolded and well lashed).[25] Gaston recommends carrying a switch about two and a half feet long for striking "one's servant or one's hound, as needed."[26] The hounds' subordination reinforces the hunt's mimesis of a social and natural hierarchy. On the other hand, as the double use of Gaston's switch suggests, the integration of hounds and hunters in the task of hunting is so thorough that humans are in some ways not distinct from, and not distinctly superior to, the hounds.

The hunt *à force* depends on cross-species communication between hounds and humans. Horn calls combine short and long notes with silences between. Twiti's treatise consistently illustrates that both hounds and human hunters are instructed by horn calls. For example, if a hart being steered toward archers passes out of the boundaries set up in advance for his path, he would "blow in this manner: a mote and then repeat it, *trorourourout, trorourourout, trorourourout* . . . to bring near me the men who are all around the hunting field, and to call back the hounds who have passed beyond the boundary."[27] From the other side, the treatises agree that the hounds' barks and bayings are meaningful to humans: "the good huntsman must know and listen for the cries and voices of his hounds, especially of the good and wise ones," Gaston recommends, so as to get information from them about what they are scenting.[28] *La Chace dou Cerf* similarly assumes that information is passed back and forth between species:

> Se tu oz c'ons chiens le destorne,
> A çaus qui ne l'auront oï
> Dois parler, se saches de fi,
> Et lor dois dire assez, non po:
> Ta ça ta ça ta ho ta ho!

> If you hear that one of your hounds has gone back to the correct track, you must speak to the hounds who have not heard him, saying to them over and over, "Ta ça ta ça ta ho ta ho!"[29]

Hunting cries are the most puzzling element in this cross-species communication system. They are curiously verbose and obscure at the same time. Twiti's treatise in English provides this cry, typical in its elements, to hounds who have gone off the correct scent and need to be started again on the scent of the chosen prey: "And if yowre houndes chase the hare or the hert and the houndes be at defauut, ye shal say in this maner then 'sohow, hossaine, hossaine, stou, ho ho sa, hossaine, ariere, hossaine, sohow.'"[30] What can this mean? "Hossaine" and "ariere" are French words that credit the hounds with verbal comprehension: first to "go back," "ariere," and more complexly, to hear in "hossaine" the information that they could be whipped if they do not go back: "hossaine" is one of the French words for that switch Gaston recommends using on servants and hounds.[31] Crediting hounds with some word recognition is reflected as well in the injunctions to use their names when speaking to them: "the lymner ay tyl his hounde be falle inne aȝein shal speke to hym callyng his name, [be it] Loiere or Bailemond or Latymere or Bemond."[32]

Alongside the French and English vocabulary that informs the hounds, and alongside addressing each one by name, two- and three-syllable collocations such as "so how," "ho ho sa," "ta ça," and "ci va" are highly characteristic of hunting cries. In some cases it is clear that these are contracted versions of meaningful phrases such as "par ci il va" (he goes this way), which becomes "ci va," and "veëz le ci aler" (see him going here), which becomes "veci" or "velci."[33] Particularly in England, such collocations seem to make sense as degenerate or corrupted Anglo-French phrases, inaccurately recalled by English speakers.[34] A second explanation for these contractions, compatible with corruption, is that "so how" and "veci" are elitist and prestigeful: precisely as they lose their general referentiality, they contribute to the mystification of elite hunting.[35] These explanations are plausible, but they cannot account for all aspects of the cries' history. First, the condensations typical of hunting cries appear not only in English but also in continental French treatises, where language capacity is not in decline. Such condensations appear even in the earliest French treatise, *La Chace dou Cerf,* suggesting that they functioned in some way prior to appearing esoteric.[36] Second, "prestigeful" as it may be to preserve French-sounding commands in English treatises, in fact the English treatises often translate the cries, or parts of them, into English versions.[37]

I propose that a different linguistic process can better account for the hunting cries' characteristics than either "corruption" or "elitism." In my view, hunting cries adapt French and English to communicate with nonspeakers. In the analogous, widely recognized registers of "baby talk" and "foreigner talk," linguists have shown that babies and foreign language speakers are credited with little or no knowledge of syntax and grammar, but with the ability to associate a few vocal sounds with meanings. Tactics for communication with them most often try simplification and repetition. Think of the tourist in Berlin saying to a taxi driver, not "we're going to the airport," but "airport, airport" hoping that simplifying the message and repeating it will do the trick. The same tactic works with dogs, indeed even better than with a German taxi driver since dogs are not expecting to hear "flughafen" in place of "airport." Augustine of Hippo noticed this canine advantage, remarking that the diversity of languages makes one more comfortable with one's own dog than with a foreigner.[38]

The hunting calls resemble "foreigner talk" and "baby talk" (or "parentese") in several ways. Like baby talk, the cries rely heavily on consonant-vowel combinations (to *ci va* and *veci* compare mummy for mother, bubba or daddy for father, tummy for stomach, wawa for water). At the same time, as for foreigner talk, the calls use a reduced repertoire of ordinary words such as "ariere" and "hossaine" in virtually syntax-free constructions, removing copulatives, pronouns, and definite articles. Speaking to a tourist inquiring "Louvre? Louvre?" a Parisian might respond not with "prenez la deuxième rue à gauche" but with "deuxième, à gauche," with gestures to help communicate the message. Equivalent simplifications abound in baby talk, such as "dolly pretty" for "the doll is pretty" and "mommy bye bye" for "mother is leaving."[39] Reinforcing foreigner talk with gestures is equivalent to the waving and touching that accompanies baby talk, meant to convey meaning with limited verbal resources. Hunting cries analogously communicate with hounds through simplification, repetition, and the unmistakable gesture of those whips.

Gaston urges the hunter to "speak to his hounds in the most lovely and gracious language that he can, which would be long and complex to write down, especially when they are tired, or are far behind, or in bad weather. . . . Huet of Vantes and the lord of Montmorency have the most handsome language and excellent sounds [*consonances*] and good voices and fine handsome ways of speaking to their hounds."[40] Gaston's term "consonances" for the "sounds" of hound talk introduces one last element that relates it to baby talk: intonation. A more accurate translation for "consonances" would be "harmonies," except that we do not often think of speech as musical. Gaston appears to say that a distinctive intonation characterizes "hound talk." A distinctive intonation also characterizes baby talk: the sing-song, slowed-down rhythm that helps infants distinguish phonemes and words more easily than if they hear ordinary adult speech patterns.[41] Possibly a similar boost to comprehension inspires what Gaston calls the "harmonies" of speech to hounds, particularly given the outdoor distances over which its words must carry.

Does the dependence of the hunt *à force* on hounds for success, and its cross-species "hound talk," undermine its mimesis of aristocratic superiority? The familiar answer would be that when Gaston and Edward praise the "noblesse of houndes," they enhance the symbolic status of canines in order to make human contact with them less abasing.[42] This is a compelling analysis, one that Howe defends for modern foxhunting.[43] Instead, I propose the reverse case: that contact with hounds, *in their status as beasts,* is compatible with and even necessary to the ritual program of celebrating aristocratic authority. In contrast to more familiar performances of authority such as a royal entry, the Court of Chivalry, or a holiday feast, ritualized hunting displays how well lordship operates both in and beyond human society—how completely the hunting party can understand and manipulate the natural world. At this point, I need to revise my earlier description of the message of ritual hunting (aristocratic control over the social and natural world) to specify that "control" entails an *informed mastery* of the natural world, not just its violent domination. The importance of knowing and understanding nature helps explain why Twiti, Gaston, and Edward would choose not only to hunt but to compose books about hunting. In these books they display their knowledge of the rules and terminology of hunting, but also their extensive knowledge of animals' habits and behaviors. In the end, it is the remarkable accessibility of hounds, not their subordinate status but the degree to which humans can communicate with them and hunt in their company, that guarantees their prominence in this social ritual about knowing and managing nature.

Contact with Hunted Animals

Another potentially abasing contact between humans and animals is with the hunted beast. Despite the elevation of the boar and hart to high symbolic status as they are pursued and killed, they bring the aristocratic party into intimate physical contact with animals. In the most curious challenge to the hierarchy that typically pertains between humans and animals, several treatises specify that huntsmen should present the droppings or "fumets" of potential animals to hunt during a breakfast assembly in the forest, so that the high-status hunters may choose which hart to pursue that day.[44] Figure 2 illustrates this moment

Figure 2. Gaston Phébus, *Livre de chasse*. Paris, Bibliothèque nationale de France, MS français 616, fol. 67. By permission of the Bibliothèque nationale de France.

from the Paris manuscript of Gaston's *Livre de chasse*: in an otherwise orderly scene (with horses enclosed, quivers stowed in the trees, high and low seating for gentle and common participants, pure white napery and decorously served dishes), a huntsman moves in from the left to spread a stag's droppings on the tablecloth like so many *amuse-bouches*. The gentle hunter seated at the far right gestures toward another set of fumets in front of him, and displays a few on his right palm.[45] This ceremonious examination of fumets at the breakfast table is a curious violation of bodily hierarchies and social prohibitions on excrement. Mary Douglas's discussion of dirt as "matter out of place" comes immediately to mind: certainly the ordinary medieval practice is to keep excrement off the dining table.[46]

Why does aristocratic hunting celebrate contact with hunted animals that would ordinarily seem debasing, even polluting? As for the hounds called "noble," the conventional answer could be that the hunted animals are symbolically elevated. A first source of elevation is the cultural distinction between wild and domestic animals. The marauding sow depicted in Chaucer's temple of Mars, "the sow munching the child right in the cradle," is a terrifying creature of chaos and violence, just as vicious as any wild boar, but not a worthy object of hunting because of her domesticity as well as her sex.[47] Wild animals are free of prior subordination to human will, suffused with a challenging alienness, metaphorically akin to a noble adversary in battle: recall that the boar should be urged to charge with the honorific title "mestre," as if he were a man of standing. In addition, as noted above, wild animals in ritual hunting are elevated through ceremonial treatment. In the wider culture, butchers have low social status because they deal with animals' dead bodies. In hunting, the exceptional formality of butchering the quarry has a literary master in Tristan, and the treatises praise it as an exemplary performance of "wodcraftez": that term, specifying the woodland location, helps distinguish cutting up boar on hunting from butchering domestic pigs in town.[48]

Elevating the hunted animal through symbolic and ceremonial treatment reinforces the hierarchizing message of hunting, but here for the last time I want to stress that elevation cannot fully explain how contact with animals is validated in this kind of hunting. At the moment of cutting up the quarry, in addition to the difference between woods and town signaled in the term "woodcrafts," the second half of the term, "crafts," carries weight as well: knowledge and expertise characterize the kind of control over nature that is being celebrated. Those fumets on the breakfast table call again for an expanded definition of the ideological program of hunting as aristocratic mastery of nature *through masterful understanding of nature.* There is a great deal to know about fumets. According to the treatises, the diet, size, sexual activity, and general health of a hart can all be read there as in a book.[49] And so fumets on the breakfast table, as bearers of occulted information, are not contaminating "matter out of place"; they have a place in expressing the authority of those who know how to parse them.

In summary, the hunt *à force* is structured so as to assert and act out the rightness of aristocratic domination in the human social hierarchy, and the completeness as well of human control over animals. The stratification inherent in this mimesis of social and natural hierarchy can mislead us into seeing the animals simply as subordinate creatures to be dominated, or at best as symbolically elevated performers in a demonstration of human supremacy. More fully accurate to the roles of the hounds, the hart, and the boar is the recognition that the ritual hunt needs to establish persistent, intimate contact with them. Of course, it also needs some of them to end up dead. But from sunup to sundown before that ending, the hunters' proximity to animals demonstrates the knowledge and understanding that make their domination of animals appear fully human: rational as well as courageous, and hierarchical in keeping with creation's hierarchy.

Notes

1. The earliest known treatise, from about 1250, uses this terminology: "fai tes millors chiens chacier / A force, car il est resons" (make your best hounds hunt *à force,* for that is best): *La Chace dou Cerf,* ed. and trans. Gunnar Tilander, Cynegetica 7 (Stockholm: Offset-Lito, 1960), lines 78–79. William Twiti's Anglo-Norman treatise advises rewarding hounds "s'il ount pris le cerf a force" (if they have taken the stag *à force*), translated in the English version of his treatise (ca. 1330) "if your houndes be bold and have slayn the hert with streynth of huntyng": *La Vénerie de Twiti: Le plus ancien traité de chasse écrit en Angleterre; la version anglaise du même traité et Craft of Venery,* ed. Gunnar Tilander, Cynegetica 2 (Uppsala: Almqvist & Wiksells, 1956), 38, 49 (hereafter cited as *VT*). Gaston provides a long account of how to hunt the hart "a force et par maistrise":

Gaston Phébus: Livre de chasse, ed. Gunnar Tilander, Cynegetica 18 (Karlshamn: Johanssons, 1971), 193 (hereafter cited as *LC*). Edward of York translates Gaston's phrase "with strengthe," and in independent passages uses both "with strengthe" and "with streng[t]h of rennyng houndes": *The Master of Game by Edward, Second Duke of York*, ed. Wm. A. and F. Baillie-Grohman (London: Ballantyne, Hanson & Co., 1904), 30, 83, 94 (hereafter cited as *MG* 1904). Because this edition is difficult to find, I cross-reference its page numbers with those of the English translation: *The Master of Game by Edward, Second Duke of York*, ed. Wm. A. and F. Baillie-Grohman (London: Chatto and Windus, 1909), 30, 148, 165 (hereafter cited as *MG* 1909). Both the transcription and the translation are faulty; James I. McNelis is preparing an edition for Universitätsverlag Winter.

2. On the literary and didactic aspects of hunting treatises, see Armand Strubel and Chantal de Saulnier, *La Poétique de la chasse au Moyen Age: Les livres de chasse du XIVe siècle* (Paris: Presses Universitaires de France, 1994); and Nadine Bordessoule, *De proies et d'ombres: Escritures de la chasse dans la littérature française du XIVe siècle* (New York: Peter Lang, 2000). Writers of treatises tend to draw on written sources as well as personal knowledge of hunting.

3. Clifford Geertz, "Deep Play: Notes on the Balinese Cockfight," in *The Interpretation of Cultures* (New York: Basic Books, 1973), 412–53; *Middle English Dictionary*, ed. Hans Kurath et al. (Ann Arbor: University of Michigan Press, 1952–); John J. MacAloon, "Olympic Games and the Theory of Spectacle in Modern Societies," in *Rite, Drama, Festival, Spectacle: Rehearsals toward a Theory of Cultural Performance*, ed. John J. MacAloon (Philadelphia: Institute for the Study of Human Issues, 1984), 241–80.

4. *LC*, 58: "Aprés le sangler le mire, et aprés le cerf la biere"; also cited by Hardouin de Fontaines-Guérin (1394), *Le Trésor de vénerie*, ed. Jérôme Pichon (Paris: Techener, 1855), 52: "pour le sengler faut le mire / Mais pour le cerf convien la bière."

5. *MG* 1904, 28; *MG* 1909, 49; *LC*, 88: "C'est une orgueilleuse et fiere beste." On the symbolic associations of hart and boar, see John Cummins, *The Hound and the Hawk: The Art of Medieval Hunting* (New York: St. Martin's, 1988), 68–83, 96–109; Marcelle Thiébaux, *The Stag of Love* (Ithaca, N.Y.: Cornell University Press, 1974); Marcelle Thiébaux, "The Mouth of the Boar as a Symbol in Medieval Literature," *Romance Philology* 22 (1968–69): 281–99.

6. *VT*, 34: "Vous devez corneer aprés la moete deus mootz. Et si vos chiens ne vienent mie a vostre volunté si hastivement cum vous vodriez, vous devez corneer quatre mootz pur hastier la gent ver vous e pur garnier la gent que le cerf est meü."

7. For more detail, see Cummins, *Hound and the Hawk*, 32–46; Marcelle Thiébaux, "The Mediaeval Chase," *Speculum* 42 (1967): 260–74; Anne Rooney, *Hunting in Middle English Literature* (Cambridge: Boydell Press, 1993).

8. *LC,* 251 ("deduit d'omme gras ou d'omme vieill ou d'un prelat"); *Les Livres du roy Modus et de la royne Ratio,* ed. Gunnar Tilander, 2 vols. (Paris: Société des anciens textes français, 1932), 1:157–71 (composed between 1354 and 1377).

9. President Theodore Roosevelt articulated the ideal of self-reliant hunting in his foreword to *MG* 1904, xii: "There was a very attractive side to the hunting of the great mediaeval lords, carried on with an elaborate equipment and stately ceremonial, especially as there was an element of danger in coming to close quarters with the quarry at bay; but after all, no form of hunting has ever surpassed in attractiveness the life of the wilderness wanderer of our own time—the man who with simple equipment, and trusting to his own qualities of head, heart, and hand, has penetrated to the uttermost regions of the earth, and single-handed slain alike the wariest and the grimmest of the creatures of the waste."

10. Sally F. Moore and Barbara G. Myerhoff, "Introduction: Secular Ritual: Forms and Meanings," in *Secular Ritual,* ed. Sally F. Moore and Barbara G. Myerhoff (Assen: Van Gorcum, 1977), 3–24 (quotation at 23).

11. My phrasing borrows from MacAloon, "Olympic Games," 253: "The sight of heretofore stoic and 'Olympian' athletes weeping under the immense symbolic weight of the victory rite is surely one of the most powerful and evocative images generated by the modern world."

12. Eve Kosofsky Sedgwick, *Touching Feeling: Affect, Pedagogy, Performativity* (Durham, N.C.: Duke University Press, 2003), 71.

13. Joyce E. Salisbury, *The Beast Within: Animals in the Middle Ages* (New York: Routledge, 1994), 45; Dorothy Yamamoto, *The Boundaries of the Human in Medieval English Literature* (Oxford: Oxford University Press, 2000), 101.

14. Cummins, *Hound and the Hawk,* 41; similarly, Strubel and Saulnier, *Poétique de la chasse,* 161: "la boucherie est anoblie par le rituel" (the butchering is ennobled by ritual).

15. *LC,* 250–51: "mes de ce parle je mal voulentiers, quar je ne devroye enseigner a prendre les bestes si n'est par noblesce et gentillesce"; traps and snares are hunting "a court deduit et vilainement."

16. *LC,* 88–92, 231–35 ("Il fuira bien de souleill levant jusques a souleill couchant," 90; "un sangler fuit bien longuement," 231). On the forty-four manuscripts of Gaston's work, see *LC,* 24–35; see also the facsimile of B.N.F. f. fr. 616: *The Hunting Book of Gaston Phébus: Manuscrit français 616, Paris, Bibliothèque nationale,* introduction by Marcel Thomas and François Avril, trans. Sarah Kane; commentary by Wilhelm Schlag (London: Harvey Miller Publishers, 1998).

17. *LC,* 233–35: "c'est belle maistrise et belle chose qui bien scet tuer un sangler de l'espee" (235).

18. Cummins, *Hound and the Hawk*, 4, 101–2; Thiébaux, "Mediaeval Chase," 261.

19. *LC*, 232; cf. *Modus et Ratio*, ed. Tilander, 1:81 (urge the boar to charge with "or cha, mestre!").

20. Quoted in R. H. C. Davis, *The Medieval Warhorse: Origin, Development and Redevelopment* (London: Thames and Hudson, 1989), 107–8.

21. *La Chace dou Cerf*, ed. Tilander, lines 23–27: "Li deduiz est si souverains / Qu'i n'est rois, ne cuens, ne Gauvains, / S'il estoit vis et bien l'amoit, / Qui plus honorez n'en ceroit."

22. James Howe, "Fox Hunting as Ritual," *American Ethnologist* 8 (1981): 278–300 (quotation at 296). On hierarchy within the huntsmen's roles, see Cummins, *Hound and the Hawk*, 172–86.

23. Moore and Myerhoff, "Introduction: Secular Ritual," 8.

24. *Modus et Ratio*, ed. Tilander, 1:58: "Dieu . . . crea deulz manieres de bestes, les unes que il apela bestes humaines et les autres furent apelees bestes muez, et furent dites muez, pour ce que il n'ont point de connoisance du Createur."

25. *MG* 1904, 97; *MG* 1909, 170. The problem of hounds changing from the original scent to a new one is a major concern in all the treatises.

26. *LC*, 193–94: "une verge qui doit avoir deux piez et demi de long. . . . on fiert de ce baston . . . a son vallet ou a un chien quant mestier est."

27. *VT*, 36: "corneer en ceste manere un moot e pus rechater trourourourout, trourourourout, trourourourout . . . de aver les gentz que sunt entour la chace a moy e rechater les chiens que sunt passez hors de bounde." Cummins discusses horn calls in *Hound and the Hawk*, 160–69.

28. *LC*, 204: "le bon veneur doit cognoistre et entendre les gueles et menees de ses chienz, espiciaument des bons et saiges"; *Modus et Ratio*, ed. Tilander, 1:41–42, makes a similar argument.

29. *La Chace dou Cerf*, ed. Tilander, lines 320–24. Gaston refers more than once to canine "language," e.g. *LC*, 134: "They hunt all day long speaking and shouting in their language and saying cruel things to the beast they hope to seize" ("ilz chascent tout le jour en parlant et en riotant en son langaige et en disant biaucoup de vilenie a la beste qu'il veut prendre").

30. *VT*, 50.

31. Gunnar Tilander, *Nouveaux essais d'étymologie cynégétique*, Cynegetica 4 (Lund: Bloms, 1957), 40–50.

32. *MG* 1904, 95; *MG* 1909, 167.

33. See Gunnar Tilander, *Mélanges d'étymologie cynégétique*, Cynegetica 5 (Lund: Bloms, 1958), 55–92. *LC*, 173, uses "par ci va par les fumees" (he is going this way by his droppings) and "vez le ci aler" (see him going here), but also uses the condensation "par cy, par cy," 146, 174; in an independent chapter Edward uses "cy va, cy va, cy va": *MG* 1904, 95; *MG* 1909, 167. Jacques de

Brézé, Grand Seneschal of Normandy, composing *La Chasse* ca. 1481–90, gives a full phrase and its condensation next to each other to represent a cry to the tracking hound: "Veez le cy aller! / La, compains, vaulcy, va avant!" (See him going here! Now, friend, *vaulcy*, go on!): *La Chasse, Les Dits du bon chien Souillard, et Les Louanges de Madame Anne de France*, ed. Gunnar Tilander, Cynegetica 6 (Lund: Bloms, 1959), 33.

34. *The Tretyse off Huntyng*, ed. Anne Rooney, Scripta: Mediaeval and Renaissance Texts and Studies 19 (Brussels: UFSAL, 1987), 32: cries given in treatises show "corrupt words of French derivation" or "pseudo-French." In a similar mode, Tilander proposes that longer cries condense to two or three syllables because they are repeated so often on the hunt; he cites other hunting work agreeing that the impulse is "abbreviation": *Mélanges d'étymologie cynégétique*, 63–65.

35. Cummins, *Hound and the Hawk*, 113: preservation of French-derived cries in English treatises is "a good example of the primacy of the sophisticated French in the hunting practices of the late Middle Ages"; Howe, "Fox Hunting," makes a similar argument on hunting terminology in general (283–84). To be sure, hunting treatises do teach esoteric terminology for animals' hides, colors, ages, fat, feet, droppings, and so on, but esoterism is not the best explanation for the hunting cries.

36. Tilander, *Mélanges d'étymologie cynégétique*, 264–79: "ta ho" in *La Chace dou Cerf* is "tahou" in *LC*, "taillaut" in Jacques de Brézé's *La Chasse*, and "tally ho" in modern English.

37. Already in the Middle English translation of Twiti's treatise, hybrid calls appear such as "here, how, here, douce, how, here": *VT*, 48. Further examples appear in *Julians Barnes Boke of Huntyng*, ed. Gunnar Tilander, Cynegetica 11 (Karlshamn: Johanssons, 1964) (first published 1486), and in *The Tretyse off Huntyng*, ed. Rooney (ca. 1400–1470).

38. Augustine, Bishop of Hippo, *The City of God against the Pagans*, trans. George E. McCracken et al., 7 vols. (Cambridge, Mass.: Harvard University Press, 1957–72), 6:148–49 (book 19, chapter 7).

39. Charles A. Ferguson, "Absence of Copula and the Notion of Simplicity: A Study of Normal Speech, Baby Talk, Foreigner Talk, and Pidgins," in *Pidginization and Creolization of Languages*, ed. Dell Hymes (Cambridge: Cambridge University Press, 1971), 141–50; Charles A. Ferguson, "Baby Talk in Six Languages," in *Language Structure and Language Use: Essays by Charles A. Ferguson*, ed. Anwar S. Dil (Stanford, Calif.: Stanford University Press, 1971), 113–33; Lila R. Gleitman, Elissa L. Newport, and Henry Gleitman, "The Current Status of the Motherese Hypothesis," *Journal of Child Language* 11 (1984): 43–79.

40. *LC*, 199: "il doit parler a ses chienz dou plus bel et plus gracieux lengaige qu'il puet, les quieulx seroyent longs et divers pour escrire, espiciaument quant ilz sont las ou ilz chascent de fort longe ou par mau temps. . . .

Huet des Vantes et le sire de Monmorenci orent de trop biaux lengaiges et trop bonnes consonances et bonnes voiz et bonnes manieres et beles de parler a leurs chienz." Like the "lord of Montmorency," the titled Anne de Beaujeu, daughter of Louis XI, "spoke constantly to the hounds" according to Jacques de Brézé, *La Chasse*, ed. Tilander, 36: "De parler aux chiens ne cessoit."

41. Erik D. Thiessen, Emily A. Hill, and Jenny R. Saffran, "Infant-Directed Speech Facilitates Word Segmentation," *Infancy* 7 (2005): 53–71.

42. *MG* 1904, 42, 44; *MG* 1909, 75–76, 88; *LC*, 106–11.

43. Howe, "Fox Hunting," 290–93; see also Garry Marvin, "Unspeakability, Inedibility, and the Structures of Pursuit in the English Foxhunt," in *Representing Animals*, ed. Nigel Rothfels (Bloomington: Indiana University Press), 139–58.

44. Droppings are presented at breakfast in *LC*, 170–72; *MG* 1904, 93; *MG* 1909, 164; Jacques de Brézé, *La Chasse*, ed. Tilander, 30–31; *Modus et Ratio*, ed. Tilander, 1:35–36.

45. The four manuscripts with most textual authority illustrate the scene in this layout, demonstrating its presence and layout in the lost presentation copy for Philippe le Hardi: Carl Nordenfalk, "Hatred, Hunting, and Love: Three Themes Relative to Some Manuscripts of Jean sans Peur," in *Studies in Late Medieval and Renaissance Painting in Honor of Millard Meiss*, ed. Irving Lavin and John Plummer, 2 vols. (New York: New York University Press, 1977), 1:331–35; plates, 2:114.

46. Mary Douglas, *Purity and Danger: An Analysis of the Concepts of Pollution and Taboo* (London: Routledge, 1966), 36 et passim.

47. Geoffrey Chaucer, *The Riverside Chaucer*, 3rd ed., ed. Larry D. Benson et al. (Boston: Houghton Mifflin, 1987), *Canterbury Tales*, A 2019: "the sowe freten the child right in the cradel."

48. Game should be cut up by one "wys vpon wodcraftez": *Sir Gawain and the Green Knight*, 2nd edition, ed. J. R. R. Tolkien and E. V. Gordon (Oxford: Clarendon, 1967), line 1605; Edward uses "woodmannys craft": *MG* 1904, 100; *MG* 1909, 176. See also François Remigereau, "Tristan 'maître de vénerie' dans la tradition anglaise et dans le roman de Thomas," *Romania* 58 (1932): 218–37.

49. *VT*, 40–41; *LC*, 62, 151–53; *MG* 1904, 18, 75; *MG* 1909, 29–30, 133–35. In Jacques de Brézé's hunt, "Des fumees eut en maincte guise / Sur la table et de lieux divers" (on the table were many kinds of fumets from various places); the experienced hunters who judge the fumets are called "les maistres congnoisseurs" (the knowledgeable masters): *La Chasse*, ed. Tilander, 28, 30–31.

CHAPTER FOUR

The (Re)Balance of Nature, ca. 1250–1350

Joel Kaye

My purpose in this essay is to provide evidence for a series of claims: that balance has a history; that between approximately 1250 and 1350 a markedly new sense of what constituted balance emerged within the discipline of scholastic natural philosophy; and that by the end of this period this new sense of the form and potentialities of balance underlay the most innovative and forward-looking university speculations on nature.[1] Over this time period, medieval natural philosophers, working primarily from an Aristotelian inheritance, developed a conception of nature as a richly complex, interconnected whole—a meta-system composed of numerous interconnected subsystems, each, in turn, composed of numerous functioning bodies and parts, moving and finding meaning in relation to each other and to the functioning whole. Central to this conception was the belief that each system and subsystem (from nature as a whole to the smallest constituent functioning bodies) maintained itself in what we would *today* term a state of "balance" or "dynamic equilibrium." I stress *today* because medieval thinkers never used the terms for balance and equilibrium in this way.

For medieval thinkers the words "balance" and "equilibrium" (*bilanx, aequilibrium*, and their cognates) rarely escaped or transcended their original ties to the common mechanical scale (Lat., *bilanx*) and the simple equalities (two balanced equal weights) that the scale was designed to measure.[2] In the medieval period the terms had not yet gained the metaphorical and mathematical breadth they enjoy today when we routinely speak of fields, systems, or multiple forces "in balance" or "in equilibrium." From the evidence of the *Oxford English Dictionary*, it is only long after the medieval period (primarily in the last two centuries) that the words "balance" and "equilibrium" have commonly come to describe a dynamic state in which multiple objects and forces are systematically ordered and integrated within a relational field.[3] Both the phrase and the concept of a "balance of nature," so common in the present day, were unknown in the period 1250–1350. The first English use of this phrase (as cited in the *OED*) occurs surprisingly late, in 1909. The first use that actually conforms to our modern sense is found in 1923, by H. G. Wells, and even then it appears in the context of science fiction.[4] And yet I want to argue that the *sense* conveyed by the phrase "balance of nature" was very much alive and active in scientific speculation, *avant la lettre*, from the ancient world through the premodern period.

With the terms "balance" and "equilibrium" expressing little more than the notion of a simple equality between two weights, and in the absence of any direct equivalents, Latin thinkers used a cluster of related words and terms centered on notions of equality and equalization (which I define here as the process of attaining or maintaining equality) to convey many (although not all) of the meanings we today attach to the word "balance." Belonging to this cluster are *medium, medietas, mediocritas, aequalitas, aequitas, aequus, aequare, aequabilis, aequivalentia, adaequatio, adaequare, proportionalitas, proportionare*, and related forms.[5] The frequency and plasticity with which these words were used indicates that the absence in the medieval period of specific terms to carry the rich meanings of our modern "balance" and "equilibrium" in no way speaks to the parallel absence of many of the senses now captured by these words.

But it is not only the meanings and uses of the *word* "balance" that have changed. This is true for countless other words. Balance, I want to argue, is different, and positing its change over time poses unique problems and questions. For one thing, in our common understanding, balance is tied to a generalized *sense*—our physical awareness of our bodies and selves within our environment—a sense that seems to exist beneath consciousness and thus resists the notion that it might be historically or culturally modified. Similarly, balance is often used to designate a larger *feeling* for how objects and spaces are or ought to be arranged, for how things properly fit and work together in the world. This last usage extends to an exceedingly wide range of subjects but is particularly central to discourses on politics, ethics, aesthetics, and the analysis of nature. It extends from profound speculations on the cosmic order down to our sense of unease when we see a picture hanging unevenly on a wall.

Recognition of the range of subjects within which the sense of balance comes into play allows us to appreciate its great importance to our psychological, intellectual, and social life, but it also tends to encourage a biological and hence essentialist understanding of it. Since we recognize that balance as an interior sense or feeling is natural to ourselves and to all humans, it is hard for us to imagine it as having a history—as developing within specific cultural contexts and as changing in form over historical time. Nevertheless, my research has led me to conclude that the sense of what constitutes balance does indeed have a history. And so too do the meanings attached to the many terms expressing equality and equalization (through which the sense of balance was conveyed in the premodern period), despite their mathematical roots and their promise of a timeless mathematical precision. While the same words (*medietas, aequalitas, aequitas, adaequatio,* etc.) continued to be used across disciplines throughout the medieval period, the definitions and processes denoted by these words changed profoundly, with a dramatic shift occurring over the period 1250 to 1350.

Although this shift in the understanding of balance has so far escaped the recognition of historians, I maintain that it profoundly altered the scope and content of speculation within numerous areas of medieval thought, including medicine, political theory, economic thought, and natural philosophy.[6] Further, I hope to show that the new model of equilibrium, which took shape between approximately 1250 and 1350, underlay and in many ways made possible the momentous shift toward what we now characterize as "modern" speculation in each of these intellectual areas and others beside them. In this essay, however, I limit my discussion to a single field of speculation within scholastic natural philosophy, intending through the use of a single case to provide an introduction of sorts to the larger project.

The project to illustrate the "rebalance of nature" between ca. 1250 and 1350 confronts three immediate methodological issues: (1) how to move from considerations of equality and processes of equalization (which were frequently named and discussed in the Middle Ages) to the larger and more diffuse apprehension of balance; (2) how to move from mathematically expressible and potentially concrete ends and processes (contained in the various terms denoting equality and equalization) to an abstract inexpressible *feeling* for how things work and are held together, which found no direct term of expression in the Middle Ages and which, even today, frequently lies beneath the level of consciousness and verbal expression; and (3) what is to be gained by making these moves?

While the named concepts of equality and equalization continue to be a focus of my investigation, I now see them as shaped by and subsumed within a deeper, historically shifting sense of balance that existed (and still exists) primarily beneath the level of consciousness and verbal expression. But this conclusion raises an even larger epistemological question central to the project of intellectual history: How do intellectualized concepts connect to or arise from the realm of wordless sense and experience? My tentative answer to this question, which I explore over the course of this essay, is that in the premodern period, before balance became a subject of analysis or discussion in itself, the wordless sense of balance was joined to voiced concepts of equality and equalization through an intermediary "model of equilibrium."

From the earliest pre-Socratic speculations on nature to those of the present day, where there is science one almost always finds, at the least, the assumption of a continuous and ordered (often cyclical) process of reciprocal interchange along with the recognition of an overarching conservation in the system of nature. And yet through the premodern period, the possible forms of balance lying at the root of this interchange were never the subject of conscious scrutiny or discussion.[7] Since the forms of balance were never verbalized or communicated from thinker to thinker, they cannot be thought of or treated as "ideas" in the normal sense. Even the word "concept" often carries too many connotations of conscious definition to be applied to them. By speaking instead of a "model" of equilibrium, I intend to bring to the fore the sensual attributes of shape and patterned motion, and (deeper still) of rhythm and anticipation, which give order, weight, and generative power to the sense of balance.

As I have come to see them, models of equilibrium are composed of interlocking assumptions, both implicit and explicit, both conscious and unconscious, that are generated in three distinct yet mutually reinforcing ways: (1) through the intellectual comprehension and internalization of forms of equalization embedded and often explicitly worked out within authoritative texts and contemporary debates (e.g., the central importance of concepts like *medium* and *equalitas* to the writings of Aristotle and Galen and their commentators); (2) through direct participation in an intellectual culture in which the model (and the assumptions underlying it), although never the subject of conscious analysis or debate, is passed, as it were, back and forth between thinkers, gaining shape, definition, and logical coherence as the generalized form is intuited and internalized; and (3) through the thinker's daily interactions with and interventions in concrete social forms of equilibrium—the institutional structures of the university, the political structures of the *civitas*, the economic structures of the marketplace, and the social structures of the living city itself—to the end of establishing personal equilibrium within historically specific social, political, and physical environments.[8] The construction of the model of equilibrium out of these three distinct elements, one primarily intellectual, one primarily intuitive, and one primarily experiential, underlies its mediating function between unconscious "sense" and conscious "science." Scattered thoughts, experiences, and intuitions of balance gain intellectual power as they find organization and form within the model.

As I envision them, models of equilibrium are culturally and historically specific. On one level, each thinker possesses or creates his or her own model, which may well be reshaped over the course of a person's life. In other words, the model has real existence and presence within the individual mind. At the same time, I want to argue that models of equilibrium are generalized and shared within particular intellectual cultures. For this reason, it is helpful at times to think of and to refer to models in the plural. Each cultural and institutional context and each generation or cohort may generate its own characteristic forms. Within these social contexts, the construction of the model is, to a large degree, a common, if unconscious, project. To the extent that individual thinkers share with others a social and political environment, an intellectual inheritance, or an institutional setting with particular institutional goals, values, and training, they are open to communally sharing (and shaping) a model of equilibrium. The University of Paris in the mid-fourteenth century, particularly the circle of natural philosophers that formed around the Arts Master Jean Buridan, is an excellent example of an intellectual culture producing a model of equilibrium that was shared to a considerable degree.

Since models possess both shape and weight, they can carry meaning, and they can convey meaning. As they were shared, they were continually open to change and refinement, not through open debate, since they were not the subjects of debate, but through a process whereby thinkers brought the equilibrium they projected onto nature ever more closely into line with the equilibrium they experienced in their physical, social, and technological environments. In short, models were (and, I believe, still are) refined in the direction of establishing "satisfying" solutions: solutions that along with fulfilling intellectual requirements also make intuitive sense; solutions that "fit" (*convenit*) and were "fitting" (*conveniens*), to use words of great currency in medieval philosophical discourse; solutions that "worked" because they were congruent with the group's sense and experience of how things worked in their world.

One of the determining characteristics of the new model of equilibrium as it took shape in the last decades of the thirteenth century is that the locus of balance shifted away from its traditional basis in the individual to a base in the complex of factors and forces that comprised the functioning system. In scholastic speculation on a wide variety of subjects—including economics, politics, medicine, and natural philosophy—thinkers moved toward a view of nature as a meta-system composed of levels of intersecting systems, each, through its functioning principles, maintaining itself in what we might today call dynamic equilibrium. Within these moving systems, questions of equality and equalization remained central and were often explicitly considered, but both the notions of what constituted equality and the means of determining it were changing.

Where formerly, equivalence was visualized as a precise match of knowable numbers or points, or as a precise static point of balance between two equal weights on a scale, now the emphasis was placed on finding proportionable "latitudes" or line ranges through which to compare and commensurate quantitative and qualitative motions within the moving system. This, in turn, pushed the mathematics of equality away from arithmetic, with its base in discrete number, toward a more fluid geometry based in ratios and the intersection of expanding and contracting lines. While through the mid-thirteenth century equilibrium had been viewed as a precondition of existence, built into each nature in the Aristotelian scheme, or instilled into every parcel of creation by a creating God, or yet again, the result of careful and conscious weighing and balancing on the part of the individual seeking it, by the end of the century equilibrium was being seen as a dynamic and depersonalized *product* of the activity taking place "naturally" within the moving system itself.

One important result of this development was that the individual part or nature was "freed" in a sense from the necessity of carrying balance within itself.[9] The *res* or individual thing within the *ordo* or system could actually be *im*balanced, or *ir*rational, or *dis*ordered in itself, and yet still find balance in the natural play of objects and forces within the functioning system. The more the functioning system emerged as the object of investigation, the more the value and meaning of the individual part was determined in relation to its shifting position and function within the systematic whole. Absolutist determinations and fixed hierarchies attached to individual natures dissolved in the face of an increasingly thorough-going relativity, as the system was reconceived as a relational field.[10]

Building on these developments, the outline of a model of equilibrium, recognizably different from any that had come before it, appears in the first quarter of the fourteenth century in the writings of the Oxford Calculators.[11] As the mathematical and philosophical concerns of the Calculators moved from Oxford to Paris in the second quarter of the century, the internal logic that bound together the model's conceptual elements continued to strengthen, and the contours of the elements themselves continued to evolve and gain definition. Among the major constituent elements recognizable by the mid-fourteenth century are the conceptualization of quantitative and qualitative increase and decrease as continuous motion along continua; the representation and measurement of qualitative change by means of graded line ranges or "latitudes"; the consequent expansion of geometry as a tool to analyze, describe, and represent the workings of nature; the refinement and increased application of relational thought and relativized determinations of value; and the acceptance of approximation and probability as ways of knowing.[12] The increasing definition of these individual conceptual elements was accompanied by an increasing tendency for them to be found acting together and logically interconnected within a vision of the systematic balanced whole.

One final word to add to the description of the models of equilibrium that emerged in this period is "proto-mechanical." We are still a long way from the self-consciously mechanical philosophy of the seventeenth and eighteenth centuries, and there are still clear residues of pre-mechanical principles and assumptions within even the most innovative speculations of this period.[13] But ever since the work of the great historian of science, Pierre Duhem, in the early twentieth century, the adjective "mechanical" has continued to be applied to aspects of fourteenth-century natural philosophy. One senses that actual mechanical devices are being used as models of form and activity in a number of the most intriguing speculations on nature from this period. In Paris, the great center of innovative natural philosophy in the fourteenth century, mechanical mills were whirling, grinding, pounding, and sawing away in every quarter of the city.[14] The Parisian natural philosopher Jean Buridan, whose writings I focus on below, clearly paid close attention to the form and functioning principles of the mechanical mill (among other mechanical devices), and he makes a number of references to them in his speculations on nature's form and function.[15]

The fourteenth century was, after all, the one in which the mechanical clock first assumed its striking physical presence in the town squares of Europe. And, not surprisingly, the clock, as a (partial) metaphor for the workings of the heavenly spheres, finds its first meaningful expression in the speculations of the natural philosopher Nicole Oresme, a close member of Buridan's circle at the University of Paris.[16] Reflections of mechanical order (and of the functioning principles of actual mechanical devices) embedded in the writings of Buridan and Oresme, in what appear to be both conscious and unconscious forms, add physical shape and weight to the model of equilibrium that undergirds their most innovative speculations.

As evidence of the new model of equilibrium that had emerged by the mid-fourteenth century, and of its immense scientific potential, I offer Jean Buridan's remarkable set of speculations in the area of scientific thought we would today term "geology." Buridan pursued these speculations in two works: his commentary on Aristotle's *De caelo et mundo*,[17] Book 2, chapters 7 and 22, and his commentary on Aristotle's *Meteorologica*,[18] Book 1, chapters 20 and 21. Of the two, his analysis in the *Questiones super tres libros Metheorum Aristotelis* is both fuller and later in time.[19] The discussion in his *Quaestiones in Aristotelis De caelo* is, however, more condensed and better suited to the purposes (and length) of this essay. I have chosen, therefore, to follow the argument as it appears in the *De caelo* and to use portions from the *Meteorologica* to supplement the discussion.

At the beginning of Book 2 of the *De caelo*, Aristotle raises the question whether the heavens can be said to have a proper right and a proper left.[20] In his discussion, Aristotle posits that someone on the other side of the earth from us would see the left and right of the heavens in a way opposite to how we see it. At this point, obviously intrigued by Aristotle's suggestion, Buridan raises a question that Aristotle had not specifically considered: whether dry and habitable land might exist on the other side of the earth, from which a viewer could actually view the heavens. He framed his question as follows: "*Utrum tota terra sit habitabilis*"—Whether the whole earth is habitable?[21] Although Buridan's question appears somewhat out of place at this point in his commentary, we know, thanks to the research of Duhem, that this question and a number of the physical and mathematical presuppositions that underlie it had a history stretching back to the first Greek commentaries on Aristotle, through the Islamic world, to the decades immediately preceding Buridan's treatment.[22]

Buridan recognizes at the opening of his question that it is commonly said (*communiter dicitur*) that one quarter of the earth's surface presently lies above water and is habitable. He then raises a question Aristotle had never considered: why would any one quarter of the earth be more likely to remain above water and habitable than any other quarter? That, in turn, raises another set of questions, some of which had been asked numerous times over the centuries: Given the spherical nature of the earth, given that according to Aristotelian physics all earth falls naturally to the earth's center, given the great abundance of water with respect to land, and assuming along with Aristotle (as Buridan clearly does) that the universe is eternal (*si mundus fuerit perpetuus, ut ponit Aristoteles*), why in the fullness of time should any portion of land whatsoever remain habitable above the water?[23]

In answer to this question, Buridan first suggests that the waters have not yet covered the whole of the earth due to the unevenness of the earth's surface and the existence of mountainous heights that are insurmountable by water. After offering this possibility he immediately argues against it, and his reasons for doing so are instructive: "For at all times, many of the higher parts of the mountains descend to the valleys, and no parts, or few ascend; thus, through an infinite time [*et sic ab infinito tempore*] these mountains ought to be wholly consumed and reduced to a sphere [beneath the waters]."[24]

There are a number of startling assumptions here. Buridan's physical world—the world on which he bases his physical speculations—is eternal and undergoing perpetual change. His sense of time, inherited from Aristotle, is vastly distant from the biblical period of six thousand years or so that medieval Christians are assumed to have believed in implicitly.[25] He consistently applies his concept of an eternal universe to his speculations on nature despite fierce resistance in his day to philosophical arguments that deny the creation of the world in time *ex nihilo*, and despite the intellectual project engaged in by numerous medieval thinkers, including good Aristotelians, to construct logical arguments against an eternal world. Clearly, the infinite extension of Buridan's time frame in his speculations on natural activity (which he shares with many of his fellow Aristotelians) makes possible a considerably deeper exploration of the logic of natural systems, and it certainly heightens the attention that must be paid to the requirements of systematic equilibrium.

As noted, Buridan carefully observed the process of erosion by which mountains are slowly and continuously being worn down by water to the sea.[26] In an earth that has existed six thousand years or so, this observation goes no further. But Buridan is thinking in Aristotelian time, not Christian time, and so he reasons that if the process of erosion continues over eternity, every mountain and, ultimately, all dry land whatsoever will eventually be washed into the sea. Moreover, if the world really is eternal, as he here assumes it is, then, he reasons, all the earth that was once above the waters has *already* been washed into the sea. Given his observations and his assumptions about systematic order, he recognizes that what needs to be explained is not the possible disappearance of dry land but its continued existence above the waters, including the continued existence of mountainous heights. "Through an infinite time, then, it would seem that the whole depth of the sea ought to be filled with the earth, thus consuming the [portion of] earth that was elevated. . . . Therefore, nothing ought to remain habitable."[27]

The continued existence of any dry land is only one of the major questions to which Buridan's natural logic has led. As he proceeds, he asks not only how any land at all could remain in the present above the waters, but how, given an eternal process, and given that every portion of dry land will eventually be washed into the sea by erosion, the *proportion* of dry land to sea could nevertheless have remained *eternally constant* at one quarter to three quarters, as he postulates that it has over the eons. Through what natural processes are the mountains and heights that gradually disappear into the sea replaced by, or as he comes to see it, perfectly balanced by, the growth of dry land and mountains at some other location on the sphere of the earth? For in an eternal universe, where erosion is perpetual, such a continuous, perfectly proportioned, and balanced replacement is necessary to explain the continued existence of a fixed proportion of dry land into the present.

To answer this question, indeed to *ask* this question, Buridan imagines the whole of earthly nature as an interconnected physical system in dynamic equilibrium.[28] He then invents an elaborate physical explanation, which, as he writes, "seems probable to me and by means of which all appearances could be perpetually saved."[29] He posits the existence of a grand, integrated, moving whole, functioning entirely on the basis of geometrical and physical principles: heat and cold cause evaporation and condensation, which in turn differentially rarify and condense earth and water, which in turn causes the earth above the waters to be lighter than the earth below, which results in a slight, eternally shifting variation between the center of the earth's weight (*centrum gravitatis*) and its center of magnitude (*centrum magnitudinis*).[30] This perpetual shifting of the two centers around each other (the play in the expanding and contracting latitude that connects them) results in a continual interchange of the relatively light with the relatively heavy.[31] As a consequence, some parts of earth are continually being raised above the circle of the waters, as other parts, *in balanced or equal measure,* are being carried beneath it.[32]

Like virtually every natural philosopher of his period, Buridan was a committed Aristotelian, and in constructing this speculation he relied on a number of Aristotelian first principles: the spherical earth, the natural tendency of the element earth to fall in a straight line toward the center of the universe, the heaviness of earth relative to water, the association of heat with rarification, the position of the earth itself at the center of the circular universe, and the assumption of the eternity of the world. In the construction of his natural system, in which there is an eternal balanced interchange between dry land and the waters of the deep, Buridan could draw on the profound sense of conservation built into the whole of Aristotle's physical thought and enunciated with particular clarity in the *De generatione et corruptione*.[33] Finally, in the realm of what can be called geology, Buridan had the example of Aristotle's *Meteorologica*, which assumes at many points the process of geological displacement. At one point in this work Aristotle employs his observations on the slow progressive building up of the Nile delta to note: "It is true that many places are now dry, that formerly were covered with water. But the opposite is true too: for if they look they will find that there are many places where the sea has invaded the land."[34] Clearly, the textual weight of Aristotle, and a sense of conservation and equilibrium internalized through the committed study of Aristotle, are evident at many points in Buridan's geological speculations.

At the same time, however, Buridan is moving here in a speculative direction not taken by Aristotle, and he is seeing possibilities and potentialities in both nature and what can only be called natural equilibrium that neither Aristotle nor the thirteenth-century Latin commentators on Aristotle were capable of seeing. Following the logic of the new equilibrium, Buridan arrives at questions and conclusions that fly in the face of foundational principles of Aristotelian physics.

As noted earlier, we can easily superimpose the form of the mechanical balance on Buridan's geological model: as one particle of dry earth falls beneath the waters in one part of the sphere of the earth, another particle, of equal measure, rises above the waters at another place; as one mountain slowly disintegrates and falls, an equal weight of earth slowly rises to form a mountain somewhere else. But Buridan's model employs not the single equality of the mechanical scale, not one active balance, but rather a near infinity of risings and fallings, covering the whole of the shifting earth over all eternity. His systematic vision of natural interchange encompasses every particle and portion of the vast globe, from the minute specks of earth he observes being carried down to the sea by mountain streams, to the formation of the mountains themselves; from the part of the earth he can see, to the opposite side of the earth he can only imagine; from the infinite past, through the present, to the infinite future. The whole system is driven by a slight but perpetual incongruence (inequality) between the earth's center of gravity and its center of magnitude. And yet through it all, a perfectly proportioned equality, which one would have to call a "dynamic" equality—the generalized *product* of a dynamic equilibrium—is maintained within the whole of the functioning system, in the midst of total replacement, over all eternity. The equalization at the heart of this model transcends the individual and individual identity: it is systematic equalization on the level of generalized mass and proportion.

In his discussion of the earth's two centers and the physical implications that follow in *De caelo*, II, 7, Buridan never uses the terms "balance" or "equilibrium," but he does make explicit use of the mechanical scale (*statera*) as a descriptive image at one point in both the commentary on *De caelo* and the commentary on *Meteorologica*. When he does so, however, it is noteworthy that he employs the scale to clarify an essential element of his argument rather than to characterize his vision as a whole. He writes:

> Now the center [of magnitude] of the earth is not the center of the universe, rather, the center of the earth's weight [*gravitas*] is the center, because the earth occupies the center of the universe

by reason of its weight not its magnitude. It balances itself [*equilibrat se*] at the center of the universe by virtue of its weight, as in the mechanical scale [*in statera*] equal weights balance equally [*equales equilibrant*] against each other, even if their magnitudes are not equal.[35]

Here the image of the mechanical balance serves to underscore the point that within the functioning system imagined by Buridan, weight and weight alone is the moving force. Measurement by the scale ignores magnitudes, forms, species, and particular natures of every sort, though they are clearly present in the substances being weighed. Apart from their weight, their particular natures are irrelevant to the functioning of the system. In the case of geological interchange, the particles of earth and water have no influence on each other beyond what can be explained by their relative weights and densities.[36] This careful reductionism points to an important aspect of the new model of equilibrium: at the same time that it represents a globalization of vision and an increase in the complexity of the functioning system, it also works by *ignoring individual natures* in its strict limitation and isolation of the active factors and causal agents involved.

Aside from this particular application of the image of the mechanical scale, Buridan does not explicitly attach the concept of equilibrium to the immensity of his "imagination."[37] But if we look at his imagination as a whole, we can see that it extends beyond a mere collection of insights (even the insights of genius), beyond the mere projection of the mechanical scale, and beyond simply playing with the rich possibilities inherent within the Aristotelian system. It begins with recognizable elements from Aristotelian physics, but there is something deeper within it that pulls and pushes the pieces into an entirely new formal arrangement, creates new ways for conceiving and understanding the process(es) of equalization, transforms the nature of both part and whole, and in the process transforms Nature itself. This deeper element, I would argue, is not a concrete, expressible idea of overarching balance (which Buridan apparently lacks), but a charged new *sense* of what constitutes balance, a new feeling for the potentialities of balance, that translates into a new *model* of how the world might work and how the myriad parts and bodies within it might tie together and function within the whole. Buridan never hints that his vision of equalization or equilibrium is different from Aristotle's or from any that came before him. Equality is equality and equilibrium is equilibrium. He never seeks to communicate his model of equilibrium in itself. Indeed, he is almost certainly unaware that he has one, and even less aware (if possible) that his thoughts are being organized and directed by one. The very depth and scope of the model, and the lack of an established vocabulary or discourse through which to express or distinguish it, obscured its recognition, even from its creator.[38]

Pierre Duhem applies the term "mechanical" repeatedly to Buridan's geological thought, even while at times recognizing the limitations of its use.[39] Ernest Moody, the first modern editor of *De caelo*, II, 7, called Buridan's framing of the problem here a "strictly mechanical explanation of a geological problem."[40] More recently, Patrick Gautier Dalché has credited Buridan with having constructed "un modele mécanique grandiose des changements de la surface terrestre."[41] The descriptor "mechanical" conveys a certain sense of what Buridan is after and achieves, and it establishes a clear marker separating Buridan's thought from that which came before it, while suggesting a direct link between his thought and the science to come. But Buridan never uses the phrase, nor, I imagine would he have recognized its meaning and its implications. Furthermore, the intellectual leap implied by the movement from non-mechanical to mechanical thinking obscures the fine gradations between Buridan's vision of nature and those that both preceded and followed his. In this case I believe that the mental image connected with the adjective "mechanical" tends to obscure the numerous other elements, all of them critical to Buridan's imagination, that combined to produce the particular model of equilibrium that unified his vision of nature.

If we do make use of the word "mechanical" in reference to Buridan's thought, we should not allow its powerful resonance to overwhelm the other notable "advances" and assumptions lying beneath his speculation. In Buridan's model of geological interchange, there is no privileged position within the system: no fixed top, bottom, or medium, no set hierarchical order. There is no essential meaning attached to the *res* itself: the same particle of earth will one day form part of habitable earth and another lie buried in the watery deep. Those parts and particles of the earth which are now above the water, including, by extension, the parts that make up his Paris, his native France, and (one would have to conclude) even the holy city of Jerusalem, will one day disappear beneath the depths.

Buridan's radical denial of meaning to geographical place—the pure relativization of space and place, quite stunning in the context of his culture yet implicit in the model he follows in the *De caelo*—is made explicit in his commentary on the *De meteorologica*.[42] There he speculates that given the progressive loss of land into the sea at one edge of the mass of dry land, and the concurrent building up of dry land from the sea at the opposite edge, it is possible that the same city (without naming any particular city) can, over an immense time, move from being the most eastern of cities to the most western of cities, as its position shifts in relation to the shifting mass of dry land and sea.[43] And as this shift occurs, the medium meridian of the land mass (*medium meridianum terre*) will shift as well in relationship to the fixed stars.[44] Unspoken, yet clearly present in this scheme, is the conclusion that all cities, and all places, will eventually be swallowed by the sea, as new ones arise on newly formed and habitable land.[45]

The individual *res*, whether particle of earth or city or continent, is subsumed in the balancing whole. The meaning held by any individual point changes continually as it moves along the intersecting arcs of the process. All has been relativized. The nature and purpose of each part shifts in relation to its purely circumstantial time and place within the moving system. The site of meaning has become the totalizing system and the equilibrium that governs it; equilibrium has become the tail that wags the dog.

Here we have a rather stark departure from an Aristotelian framework, in which the concept of nature as an overarching system and process exists alongside an insistence that each of its parts possesses its own inherent nature and its own proper end.[46] In the Aristotelian system it is the irreducible "nature" of the part that determines its place within the whole, connecting and ordering it within the larger functioning system of a purposeful Nature. Notions of hierarchy, ontological grading, and individuated purpose are central to Aristotelian thought, and these aspects were fastened upon by scholastic thinkers of the thirteenth century because of their central importance within other authoritative sources of medieval culture. But meaning and purpose find little or no place within Buridan's model of geological equilibrium: it is, instead, governed by geometrical and physical necessity, driven by its own internal logic, and held together by a new sense of the possibilities and potentialities of balance when applied to the workings of nature.

There were authoritatively acceptable answers available (in both an Aristotelian and Christian sense) to the question Buridan posed of why, given the observation of erosion within an eternal universe, the waters of the sea had not completely covered the earth. One could use the fact that dry land and mountains still persist in the present as proof that the world is not eternal but is rather created in time, and a limited time at that, thereby satisfying the requirements of religious doctrine. Or one could argue, along with the astrologers, that the constant application of heavenly virtues keeps the earth in its most perfect order—the order that is found in the present. Or one could argue from an Aristotelian final cause: Nature wills the continued existence of land above the waters because it is the proper end (nature) of the earth to be habitable and the abode of animals and men. This last was the most prevalent of the philosophical solutions offered before Buridan, although there were proponents of all three.[47] Buridan rejects all of them. He does not even mention the first argument based on the limited created time of Christian doctrine, despite clear openings to do so. His reasoning (sometimes implicitly and sometimes explicitly) undermines the astrological argument in virtually every particular. And he raises the possibility of explaining the phenomena through Aristotelian final cause only to dismiss it lightly.[48] In short, he abandons fixed natures, fixed meanings, fixed places, fixed times, fixed hierarchies, and mind-directed "purpose" as he follows the logic of the moving system in equilibrium.

As further evidence of the radical difference between the Buridanian and Aristotelian models of natural equilibrium, I conclude with a discussion of Buridan's commentary on the *De caelo*, Book 2, question 22, where he asks: "Whether the earth remains fixed and motionless at the center of the universe?"[49] Buridan should not have to ask this question. Aristotle was in full agreement with revelation on this point, asserting and demonstrating at many points throughout his writings that the place of the earth was fixed at the center of the universe. To contradict Aristotle here would be to contradict one of his most foundational principles, yet this is the path that Buridan follows.

Question 22 of Buridan's commentary on the *De caelo*, Book 2, is a landmark in the history of science. It begins with Buridan's arguments for the possible (probable?) daily rotation of the earth to explain the apparent daily rotation of the sun and the other heavenly spheres around the earth. It utilizes arguments based on the relativity of motion and perception that would be used again—to similar purpose—by Copernicus in his *De revolutionibus orbium coelestium*, nearly two centuries later. Much scholarly attention has been paid to this speculation. Here I intend only to suggest that the model of equilibrium we have seen directing Buridan's geological thought is in most respects the same model that underlies his revolutionary speculations on the diurnal rotation of the earth, except that in the latter case, relative perspective substitutes for relative weight as the active factor. Buridan himself demonstrates this connection when, toward the end of question 22, he seamlessly shifts his speculation on the earth's possible motion from the realm of astronomy and perspective back to the realm of geology and weight.

Buridan begins with the Aristotelian notion that all weight is drawn naturally to the center of the universe (*medium mundi*), conceived as a point at the absolute center of the concentric planetary spheres. He then argues that in order for the earth itself (*tota terra*) to remain fixed and motionless at this same center, the center of the earth's magnitude must correspond precisely to the *medium mundi*. But what, then, is he to make of his earlier model of geological interchange, which is predicated on a disjunction between the earth's center of magnitude and its center of gravity?[50] If the geological process he outlined in his earlier question is not merely speculative but actually *probable*, as Buridan maintained it was, then as mountains continually sink at one place and rise at another, the weight of the earth must continually shift around its center, however small this weight shift might be in relation to the overall weight of the earth, and however small the resulting magnitude of the shift might be in relation to the earth's overall size.[51] Buridan writes:

> And by this another doubt is solved, that is, whether the earth is sometimes moved according to its whole in a straight line. And we can answer in the affirmative.... [As the weight of the earth shifts due to the interchange of dry land and water and the rise

and decay of mountains,] that which has newly become the center of gravity is moved so that it will coincide with the center of the universe, and that point which was the center of gravity before ascends and recedes. . . .[52]

In short, *tota terra*, the whole earth, is constantly undergoing minute rectilinear motions around the center of the universe as the surface of the earth is systematically transformed and the weight shifts accordingly. Here, the model of equilibrium has been worked out so exquisitely, and has been accorded such great intellectual weight, that it has become capable of moving the earth itself—and this despite the enormous counterweight of both scriptural and Aristotelian authority.

The "mechanical" aspects of this model of equilibrium are unmistakable—the motion of the earth here can be seen as a mechanical *product* of the system in dynamic equilibrium, an aggregate product which is now allowed the logical weight to overthrow deeply held normative assertions. Well, not quite. The last paragraph of this question frequently does not receive the attention it requires.[53] Here, Buridan explicitly renounces the mechanical motion whose logic he has just asserted. He does so, he tells us, because an earth that moves, however minutely about the center of the universe, would play havoc with the a priori (and strongly held) Aristotelian assumption that the heavenly spheres that carry the stars and planets are perfectly circular and move in perfect circles. How can a perfect circle, granted physical reality, have a moving center? The two physical assumptions clearly oppose each other, and nature, to Buridan's mind, cannot be imagined to work in this way.[54]

Bernard Ribémont explains Buridan's reversal by maintaining that his argument for the earth's motion was merely an argument from probability or "secundum imaginationem," rather than a proper physical demonstration.[55] This seems to be contradicted by the weight and probability Buridan allows this speculation. But even if Ribémont is correct, what does it say about Buridan as a mechanical thinker? In his commentary to Aristotle's *Physics* I, 16, Buridan provides an argument *against* the earth's moving in response to geological change, and his argument here is based, once again, on well-considered "mechanical" principles—the small motion caused by the displacement of the earth's two centers would be resisted by the great mass of the earth tending toward the center of the world.⁵⁶ Here, in contrast to *De caelo*, II, 22, the mechanical argument appears to have been allotted the weight of demonstration. And in his final visit to this question in his commentary on the *Meteorologica*, Buridan seems simply to assume (once again) that the earth itself, in its totality, will move in response to the shifting weight caused by the eternal interchange of dry land and sea.⁵⁷

So what do we have here? Is Buridan's thought mechanical, non-mechanical, pre-mechanical, proto-mechanical, or semi-mechanical? Or are these descriptions simply too gross and too dichotomous to provide an accurate sense of the shape and logic of Buridan's thought? At best, they present only a rough basis for comparing Buridan's conceptualization of nature to that of other thinkers, or for comparing the conceptualization shared by the Parisian natural philosophers—Buridan, Oresme, Albert of Saxony, and others—to that found in other intellectual cultures that either preceded or followed them. An adjective such as "mechanical," or totalizing descriptors of whatever sort, provides almost no way of marking directions and continuities in the development of thought over a succession of thinkers, groups, centuries, or cultures.

I suggest instead that a focus on elucidating models of equilibrium—paying close attention to the evolution of their elements, the connections and forces they assume, the power they are allowed in determining the direction of particular insights, their organizing logic, and their overarching form—has the potential to serve as a considerably more flexible instrument of historical analysis and comparison, one capable of mapping continuities and change.

But how is one to define the model's form when it is never explicitly discussed or described by the thinking subjects; when, in all probability, it was never consciously grasped by the subjects who employed it; when it is constructed from wordless sense and feeling as much as from textual influence and inheritance? Here I suggest that the very implicitness and unconsciousness of the model might render it a more fruitful subject for historical inquiry than intellectual contents that are fully conscious and made explicit in texts. Granted the difficulties of recognizing the components and form of a model that exists to a degree beneath the level of consciousness, still, the fact that it is not consciously *presented* by the thinker, not colored and tailored to fit some set of normative ideals, allows it considerable advantages as a marker of thought. This is even more the case if, as I suspect, models of equilibrium are present in some form in virtually all speculation (including pre- or nonliterate speculation) that can be characterized as scientific. One of the great advantages in bringing models of equilibrium to the center of a history of science is the potential it creates for comparative studies of historical development between and across different disciplines, periods, and cultures. As for assessing the place and power of the model of equilibrium in the process of scientific discovery, we need only appreciate the remarkably forward-looking insights it permitted Buridan in this single area of geological speculation.

Between 1250 and 1350 nature was radically reconceived within the discipline of scholastic natural philosophy. In a previous work, I offered an outline of this reconceptualization: the shift from a static world of discrete points to a world of expanding, contracting, and intersecting lines; the corresponding shift from arithmetic to geometry as the primary mathematical tool of analysis and description; the expansion of focus from the particular to the common and from the individual nature to the totality of the functioning system; the increased acceptance of approximation and probability as forms of legitimate knowledge; the replacement of a world of fixed and inherent values with a relational world in which values were determined relative to changing perspectives and conditions.[58] Each of these new directions proved to be of great importance to the future of scientific thought.

I now believe that all of these components of reconceptualization, and more, were joined and held together within an overarching yet evolving sense of how natural systems achieve and maintain themselves in balance. And I see, too, that this sense was, in turn, grounded and given intellectual shape and weight within the medium of an evolving model of equilibrium. Each of the components of reconceptualization enumerated in the paragraph above can be observed in the model underlying and unifying Buridan's geological speculations. But the particular model that connected and integrated the elements of reconceptualization for Buridan was limited neither to geology nor to Buridan himself: it was shared and shaped by a number of thinkers attached to Buridan's circle at Paris and applied by them to the analysis of systematic activity over a wide range of natural phenomena.

The new model *redefined the bounds of possibility* for the conscious conceptualization of forms of equality and equalization in nature. Due to the continued centrality of these forms to scientific discourse, from the ancient world through the medieval period, the recognition of new potentialities in the realm of equalization in turn underlay the most striking, innovative, and forward-looking speculations in the realm of natural philosophy. In short, the rebalance of nature that occurred between 1250 and 1350 was not a mere detail in the story of scientific development in this period: to a considerable degree it was the story.

Notes

I wish to thank Caroline Walker Bynum for her careful and most helpful reading of a draft of this essay. My thanks as well to the Medieval Studies Program at Ohio State University and particularly to Barbara Hanawalt and Lisa Kiser for inviting me to participate in this discussion of nature; and to the Institute for Advanced Study, School of Historical Studies, for providing me with the setting and the time to prepare the first draft of this essay; and to the National Endowment for the Humanities for supporting my year at the Institute.

1. These dates are approximate at best. The first coincides both with the early flowering of the Aristotelian commentary tradition and with the regularization of institutional structures for the study of natural philosophy at the University of Paris. The second represents a moment, approximately a century later, when Jean Buridan, the great progenitor of the Parisian school of natural philosophy in the fourteenth century, is at the height of his powers. Buridan's speculations are central to the subject of this essay.

2. As far as I have been able to tell, this holds true in the ancient Greek and Latin traditions as well. One rather rare exception, cited in Lewis and Short, *A Latin Dictionary* (Oxford, 1987), 57, col. 1, is Cicero's use of *aequilibritas* in *De deorum natura*, 1, 39, 109, to translate Epicurus's *isonomia* (equality), implying here the more complex sense of an equal distribution of the multiple powers of nature. The same exception is cited in Du Cange, *Glossarium mediae et infimae Latinitas*, vol. 1, col. 1008.

3. The phrase "balance of fortune," circa 1320, is the earliest recorded metaphorical use of balance cited in the *OED*, perhaps because of the convenient linkage between the rise and fall of the wheel of fortune (a popular image at the time) and the rise and fall of mechanical scales. The escapement mechanism that regulates the motions of the mechanical clock was first designated as a "balance" in English in 1660, and with "balance" having escaped

the simple duality of the mechanical scale, the phrase "balance of power" appears soon after.

4. Even the common phrase "sense of balance," applied to our physical sense of our body's dynamic relationship to its environment—a sense which, because it is embodied, we might imagine always existed—appears only in the nineteenth century, with its first use credited to Dickens. Searches in French, Spanish, Italian (all of which, like English, derive the word "balance" from the Latin term for the mechanical scale, *bilanx*), and German (*Gleichgewicht*) indicate a similar pattern of slow metaphorical expansion.

5. In ancient Greek, *meson* (middle), *mesotês* (medium), *isonomia* (equality—see note 2 above), and *symmetria* perform the same function. The situation in Latin is reflected in Du Cange, *Glossarium mediae et infimae Latinitas*, where the meanings allotted to *aequilibrium* and all of its cognates (*aequilibratio, aequilibritas*, etc.) occupy less than one quarter of a column (vol. 1, col. 1008), while for comparison, the meanings attached to *aequalis* occupy 6 columns, *aequitas* receives 8 columns, and *aequus* 18 columns. *Bilanx* is allotted a mere 6 lines (vol. 2, col. 1985), with no expansion of meaning beyond the mechanical scale.

6. In my larger, book-length project, of which this essay forms a part, I assess the impact of the transformation of balance in each of these areas.

7. Aristotle's treatment of justice in *Ethics* V, 5, is the closest that he comes to an abstract discussion of forms of equalization, and as such it assumes an important place in later scholastic speculation. For a premodern recognition of the relationship between forms of balance and the structuring of intellectual discourse, see Al-Ghazali, *The Just Balance* [Al-Quistas Al-Mustaqim], trans. D. P. Brewer (Lahore: Sh. Muhammad Ashraf, 1978). I am not including in this essay the history of writings on statics per se (particularly those associated with Archimedes) as they were read, translated, and commented upon in Greek, Arabic, and Latin. The nearly complete works of Archimedes were translated from the Greek into Latin by William of Moerbeke in 1269. The standard historical work on the Archimedean tradition through the thirteenth century, which includes an edition of the Latin text of Moerbeke's translations, is Marshall Clagett, *Archimedes in the Middle Ages*, 3 vols. (Philadelphia: American Philosophical Society, 1964–76).

8. Consideration of this last component forms a significant part of my larger project but lies beyond the scope of this essay. My *Economy and Nature in the Fourteenth Century: Money, Market Exchange, and the Emergence of Scientific Thought* (Cambridge: Cambridge University Press, 1998) devotes considerable attention to analyzing the impact of social contexts on scientific speculation.

9. The separation of the individual *res* from its supposed "dispositions" and the disciplined application of the "principle of economy," both of which are associated with the philosophy of William of Ockham and with nominalism itself, were among the great shared projects of fourteenth-century natural

philosophy. The parallel "freeing"of the moving *res* from carrying motion as an internal disposition, one of the major programs of Ockhamist physics, has been the subject of much historical discussion. See for example, Joel Biard, "Le statut du mouvement dans la philosophie naturelle buridanienne," in *La nouvelle physique du XIVe siècle*, ed. Stefano Caroti and Pierre Souffrin (Florence: Olschki, 1997), 141–59.

10. The use of the term "relativity" here and elsewhere in this essay is not intended to reflect modern scientific definitions of the term but rather to underscore the abandonment of fixed, inherent, and absolute values and meanings in favor of those that are determined relative to position and place within the moving systematic whole.

11. For a description of the overall project of the Calculators, see Edith Sylla, "Medieval Quantification of Qualities: The Merton School," *Archive for History of Exact Sciences* 8 (1971): 7–39; Sylla, "The Oxford Calculators," in *The Cambridge History of Later Medieval Philosophy*, ed. Norman Kretzman, Anthony Kenny, and Jan Pinborg (Cambridge: Cambridge University Press, 1982), 541–63.

12. These and related intellectual developments and their place in the reconceptualization of nature are considered at many points within my *Economy and Nature in the Fourteenth Century*.

13. This point is made particularly well by Marshall Clagett, "Nicole Oresme and Medieval Scientific Thought," *Proceedings of the American Philosophical Society* 108 (1964): 298–310, esp. 300–302.

14. For visual evidence of the strong presence of mechanical mills in the Parisian landscape in this period, see Virginia Wylie Egbert, *On the Bridges of Mediaeval Paris: A Record of Early Fourteenth-Century Life* (Princeton: Princeton University Press, 1974).

15. See, for example, the place that the observation of millworks plays in Buridan's innovative and influential speculations on impetus as an explanation for the acceleration of falling bodies: *Acutissimi philosophi reverendi Magistri Johannis Buridani subtillissime questiones super octo phisicorum libros Aristotelis* . . . (Paris, 1509; rpt. Frankfurt: Minerva, 1964), bk. 8, q. 12, fol. 120rb–121rb. This question has been translated into English by Marshall Clagett, *The Science of Mechanics in the Middle Ages* (Madison: University of Wisconsin Press, 1959), 532–38.

16. For a concise summary of the relationship between Oresme and Buridan at the University of Paris, see William Courtenay, "The Early Career of Nicole Oresme," *Isis* 91 (2000): 542–48, esp. 548 n. 20. For Oresme's scientific use of the clock metaphor, see, for example, his commentary on Aristotle's *De caelo*, II, 2, in *Le Livre du ciel et du monde*, ed. and trans. Albert Menut and Alexander Denomy (Madison: University of Wisconsin Press, 1968), 288–89. Clagett ("Nicole Oresme and Medieval Scientific Thought," 300–302) notes the distinctions between a truly "mechanical" solution to the motion of the spheres and that offered here by Oresme. Not surprisingly, though, the clock

metaphor underlies a number of Oresme's most important and forward-looking speculations.

17. Buridan's questions and commentary on Aristotle's *De caelo* can be found in two modern editions: *Iohannis Buridani Quaestiones super libris quattuor de caelo et mundo*, ed. Ernest A. Moody (Cambridge, Mass.: The Mediaeval Academy of America, 1942), and *Joannis Buridani Expositio et Quaestiones in Aristotelis De caelo*, ed. Benoît Patar (Louvain: Éditions Peeters, 1996). I follow the Patar edition.

18. This work has been edited by Sylvie Bages, *Les Questiones super tres libros Metheorum Aristotelis de Jean Buridan: Étude suivie de l'édition du livre I* (Thèse de l'Ecole de Chartes, 1986). Book 1, qq. 20 and 21 occupy pp. 288–316.

19. I follow Patar's conclusion that Buridan's commentary on *De caelo* precedes that of his commentary on the *Meteorologica*. The particular dating (the first commentary to 1328–30 and the second to 1352) is less certain. On this, see Patar, ed., *Quaestiones in Aristotelis De caelo*, "Introduction," 19, 116–17.

20. Aristotle, *De caelo*, II, 2.

21. *Quaestiones in Aristotelis De caelo*, ed. Patar, bk. 2, q. 7, 410–17. (Abbreviated hereafter as *Quest. De caelo*.) This quaestio has been partially translated (from the Moody edition) by Edward Grant in his *A Source Book in Medieval Science*, ed. Edward Grant (Cambridge, Mass.: Harvard University Press, 1974), 621–24. I use the Grant translation where possible. See also Ernest A. Moody, "John Buridan on the Habitability of the Earth," *Speculum* 16 (1941): 415–25; Bernard Ribémont, "Mais où est donc le centre de la terre," in *Terres médiévales*, ed. Bernard Ribémont (Paris: Editions Klincksieck, 1993), 261–76.

22. Duhem's remarkable study, completed in the first decade and a half of the twentieth century, is still the best and most thorough analysis both of the history of geological speculation through the fourteenth century and of the extraordinary place occupied by fourteenth-century natural philosophy, and particularly the thought of Buridan, in that history. Pierre Duhem, *Le système du monde*, vol. 9 (Paris: Hermann, 1958), 79–323.

23. *Quest. De caelo*, II, 7, ed. Patar, 410. Duhem (*Système*, vol. 9, 79–170) considers the history of this observation. See especially 102, 120–22.

24. *Quest. De caelo*, II, 7, ed. Patar, 410: "quia omni tempore partes superiores ex montibus descendunt multae ad valles, et nullae vel paucae ascendunt; et sic ab infinito tempore illi montes deberent esse toti consumpti et reducti ad planitiem."

25. Buridan is far from alone in this. He shares this sense of an eternal universe and its unproblematized application to his physical speculations with many fellow Aristotelians, dating back to the twelfth century, and it appears as an a priori assumption in numerous other of his physical speculations.

26. *Quest. De caelo*, II, 7, ed. Patar, 410: "omni tempore multae partes istius terrae altioris portantur cum fluviis in profundum maris . . ." Buridan pursues

the physical implications of eternal erosion, which he encounters both in mountain streams and in the rivers of the plains, with an even greater observational precision in his commentary on Aristotle's *Meteorologica*, ed. Bages, especially bk. 1, qq. 20 and 21.

27. Grant, *Source Book*, 621–22; *Quest. De caelo*, II, 7, ed. Patar, 411: "Ideo videtur quod ab infinito tempore tota profunditas maris deberet esse repleta terra, et haec elevatio terrae deberet esse consumpta; et sic aqua naturaliter deberet totam terram circumdare, nec deberent esse aliquae elevationes discoopertae."

28. Duhem, I think justifiably, frames his entire discussion of Buridan's geology around the concept of equilibrium, entitling his section on the subject in *Système*, vol. 9, "L'Équilibre de la terre et des mers."

29. *Quest. De caelo*, II, 7, ed. Patar, 416: "quae videtur mihi probabilis, et per quam perpetuo salvarentur omnia apparentia . . ."

30. Duhem traces the considerable history of the idea that there might exist a disjunction between the earth's *centrum gravitatis* and its *centrum magnitudinis*, going back to Alexander of Aphrodisias (*Système*, vol. 9, 81), and he illustrates the sharpening of this speculation in the writings of two of Buridan's younger contemporaries at Paris, Nicole Oresme and Albert of Saxony (*Système*, vol. 9, 202–18).

31. *Quest. De caelo*, II, 7, ed. Patar, 416: "Et ita apparet quod aliud est centrum magnitudinis terrae, et aliud est centrum gravitatis eius, nam centrum gravitatis est ubi tanta est gravitas ex una parte sicut ex altera, et hoc non est in medio magnitudinis, ut dictum est. Modum ultra, quia terra per suam gravitatem tendit ad medium mundi, ideo centrum gravitatis terrae sit in centro mundi, et non centrum suae magnitudinis, propter quod terra ex una parte est elevata supra aquam, et ex alia parte est tota sub aqua."

32. Buridan's assertion of the disjunction of the earth's two centers and his speculations on the geological implications of this disjunction are more fully developed in his commentary on the *Meteorologica*, bk. 1, q. 21, conclusions 2 and 3, ed. Bages, 308–9.

33. Buridan does not directly reference the *De generatione* as a source for the cycle of creation and destruction he is imagining here, but his near contemporary at the University of Paris, Nicole Oresme, does when he speculates on the question of cyclical geological displacement. On this, see Oresme, *Le Livre du ciel*, bk. 2, q. 31, ed. Menut and Denomy, 570: "je suppose que les elemens naturelment peuent, selon leurs [parties], crestre et appeticier par generacion et corrupcion, et ce suppose Aristote . . . ou livre de Generacion et corrupcion et en pluseurs autres lieus."

34. Aristotle, *Meteorologica*, bk. 1, q. 14. This whole question is a marvel of observation and reasoning, much of which would have been instructive for Buridan.

35. *Questiones super tres libros Metheorum*, I, 21, ed. Bages, 309: "Sed centrum terre non est centrum mundi; ymmo centrum eius gravitatis est centrum mundi quia terra non ratione sue magnitudinis sed ratione sue gravitatis tenet locum medium mundi. Ideo secundum suam gravitatem equilibrat se ad centrum mundi sicut in statera gravitates equales equilibrant se adinvicem, licet magnitudines non sint equales." See also *Quest. De caelo*, II, 22, ed. Patar, 505–6.

36. In *De generatione and corruptione* II, 4, Aristotle had assumed the possibility of "reciprocal transformation" or actual "conversion" between the elements of earth and water, but this possibility is utterly excluded by Buridan in the imagination of his working system.

37. Lynn Thorndike has found a geological argument, clearly later than Buridan's and based directly on his, in which a direct (if still limited) link is made between the modeling of geological replacement and the equilibrium of the mechanical scale. In order to explain why all earth has not disappeared beneath the waters, the author of BN 6752 (fol. 159r–v) speculates: "The heavier half of the sphere sinks the deeper below the surface of the water, while a portion of the lighter hemisphere projects out of the water, just as the heavy scale of the balances falls and the lighter rises" (Thorndike's translation). On this, see Lynn Thorndike, *A History of Magic and Experimental Science*, vol. 3 (New York: Columbia University Press, 1934), 568–84, esp. 580. New research has established Thorndike's anonymous author as the Augustinian Jacques Legrand. On this, see E. Beltran, "Jacques Legrand OESA: Sa vie et son oeuvre," *Augustiniana* 24 (1974): 395 ff.; Patrick Gautier Dalché, "L'influence de Jean Buridan: L'habitabilité de la terre selon Dominicus de Clavisio," in *Comprendre et maîtriser la nature au Moyen Age: Mélanges d'histoire des sciences offerts à Guy Beaujouan* (Geneva: Librairie Droz, 1994), 101–15, esp. 101.

38. The immensity of scope is well represented in *Questiones super tres libros Metheorum*, I, 21, conclusion 10, ed. Bages, 314: "possibile est quod in terra que nunc est discooperta generabuntur de novo montes alti versus orientem et corrumpentur aliqui magni ad occidentem, ex quibus contingit fluvios augeri et multiplicari ad orientem et deficere vel diminui ad accidentem et e contrario, et quod mare redundet ex fluviis ibi accedentibus."

39. Duhem, *Système*, vol. 9, 202.

40. Moody, "John Buridan," 420.

41. Dalché, "L'influence," 108.

42. The denial here of meaning to place is yet another element in Buridan's thought that takes shape over the fourteenth century, and it serves as yet another marker separating the model of equilibrium achieved at Paris in the mid-fourteenth century from those created by thirteenth-century Aristotelian philosophers. It also serves to separate Buridan's thought from that of contemporary astrological thought.

43. *Questiones super tres libros Metheorum*, I, 21, ed. Bages, 316: "Tertia decima conclusio est quod eamdem civitatem possibile est fieri magis orientalem quam ante esset vel magis occidentalem, quia magis orientalis dicitur civitas ex eo quod est propinquior magno mari versus orientem, etc. Fiet autem propinquior si mare ex illo latere augetur et remotior si diminuatur." Although Buridan does not state so explicitly here, since he has shown that the process is both continuous and eternal, the land beneath the city and the city itself must also eventually crumble and disappear beneath the waters.

44. *Questiones super tres libros Metheorum*, I, 21, ed. Bages, 316: "Unde, secundum quod magnum mare circuit ipsam terram, oportet mutare medium meridianum terre habitabilis in ordine ad celum ymaginatum quiescens."

45. In a number of sections of his commentary on *Meteorologica* I, 20, Buridan closely observes the changing coastline of the Mediterranean, noting the building up of certain delta islands and the erosion of other parts of the coast into the sea.

46. There is a similarly radical departure from any physics grounded in notions of microcosm/macrocosm.

47. Duhem (*Système*, vol. 9, 79–170) stresses the strength of the argument from final cause that existed before Buridan in both the Islamic and Christian Aristotelian tradition. On final cause arguments, see also Ribémont, "Mais où est donc le centre," 262. Dalché ("L'influence," 106) remarks on their persistence in certain "geological" speculations that followed Buridan's in the fourteenth century.

48. *Quest. De caelo*, II, 7, ed. Patar, 415: "quod non est conveniens dicere volentibus tenere perpetuitem mundi in statu prospero animalibus et plantis sicut nunc est."

49. *Quest. De caelo*, II, 22, ed. Patar, 500–508: "Utrum terra semper quiescat in medio mundi." This question has been translated by Marshall Clagett and appears in his *Science of Mechanics in the Middle Ages*, 594–99.

50. Buridan restates this argument in *Quest. De caelo*, II, 22, ed. Patar, 505–6. To emphasize that this is an argument driven by the factor of weight, and that the governing equilibrium is centered on weight alone, he adds the example of the mechanical scale, which, he emphasizes, works by balancing equal weights not equal magnitudes: "si in statera ex una parte ponatur lapis et ex alia parte lana, lana erit valde maioris magnitudinis." Decades later when he returned to the question of geological replacement and the motion of the earth in his commentary on the *Meteorologica*, he used the same example of the mechanical scale to underscore the same point.

51. For a similar, considerably earlier speculation on this possibility, albeit one lacking much of the logical scaffolding Buridan provides, see Al-Biruni, *The Determination of the Coordinates of Positions for the Correction of Distances between Cities*, ed. and trans. Jamil Ali (Beirut: American University of Beirut, 1967), 17.

Here, I would argue, is a case where the comparison of the physical speculations of these two thinkers, so distant in time and culture, in terms of the particular form of their underlying models of equilibrium, could be more historically revealing than a comparison based only on the similarities and differences of their particular insights.

52. *Quest. De caelo*, II, 22, ed. Patar, 507: "Et per hoc solvitur alia dubitatio, scilicet utrum terra aliquando moveatur secundum se totam motu recto. Et possumus dicere quod sic. . . . illud quod de novo factum est medium gravitatis movetur ut sit medium mundi, et illud quod ante erat medium gravitatis, ascendit et recedit . . ."

53. For example, it is left out of Clagett's translation of the question in his *Science of Mechanics in the Middle Ages*, and consequently, it is also missing in Grant's reproduction of the Clagett translation in his *A Source Book in Medieval Science*. Ribémont, however, pays close attention to this paragraph in "Mais où est donc le centre," 266–68.

54. *Quest. De caelo*, II, 22, ed. Patar, 507: "ideo frustra moveretur si moveretur; et nihil est ponendum frustra in natura. Ergo ponendum est quod non moveatur."

55. Ribémont, "Mais où est donc le centre," 267.

56. Ibid.

57. *Questiones super tres libros Metheorum*, I, 21, ed. Bages, 310–11: "Quinta conclusio sequitur quod necesse est continue semper vel aliquando terram totam simul moveri. . . . ergo falsum est quod totalis terra non moveatur."

58. Kaye, *Economy and Nature in the Fourteenth Century*, 1 and passim.

CHAPTER FIVE

Collecting Nature and Art
Artisans and Knowledge in the Kunstkammer

Pamela H. Smith

In his 1565 treatise on how a collection should be formed, Samuel Quiccheberg (1529–67) extolled collecting as "a first philosophy":

> I must explain that the invention of the first philosophy [of collecting], as it calls itself, is a novelty in the whole of Europe; it has brought about the certainty of all scientific areas as well as the most complete methods; like the opening of the doors of wisdom, it has produced the greatest use and godly clarity in the sciences.[1]

Quiccheberg titled his treatise "Inscriptions or Titles for an Ample and all-encompassing Theater which includes particulars of the whole creation and outstanding images, or . . . a storehouse of artful and wonderful things . . . which will bring about a new knowledge of things and admirable prudence [cognitio rerum et prudentia admiranda], or wisdom in statecraft." I do not think that Quiccheberg's claims that collections would bring "a new knowledge of things" and "wisdom in statecraft" were hyperbolic. Although Quiccheberg's *Inscriptiones vel*

Tituli Theatri Amplissimi has often been seen as a set of instructions for displaying the wealth and power of a prince or of representing the plenitude of the cosmos, I argue in this essay that collecting was part of a new conception of philosophy that viewed knowledge as active, productive, and based on nature. This dimension of Quiccheberg's plan for a *Kunstkammer* emerges in particular relief when compared with the work of the Nuremberg goldsmith Wenzel Jamnitzer (1510–85), his contemporary and a collector himself.

The emergence of collections of antiquities, natural objects, and works of art in the sixteenth century, as well as their significance for the study of nature, has been a familiar subject since the 1980s and 1990s, when Oliver Impey and Arthur MacGregor's *The Origins of Museums* signaled the onset of the serious study of art and curiosity chambers—*Kunst- und Wunderkammern*—in early modern Europe. Since then, the work of Thomas DaCosta Kaufmann, Horst Bredekamp, Paula Findlen, Martin Kemp, and numerous others has illuminated many aspects of this important study for the understanding of nature in the Renaissance.[2] Scholars have noted the cluster of secular princely (as distinguished from Church) collections founded in the mid-sixteenth century. August I, Elector of Saxony, established a *Kunstkammer* in 1560, the same year that Archduke Ferdinand II began assembling one at Schloss Ambras near Innsbruck.[3] It is clear, however, from Quiccheberg's comments about his ideas and models for collecting that some of the first secular collections in the north were probably those of the Fugger family, the long-distance merchants based in Augsburg, for whom Jacopo da Strada was already collecting in the 1540s.[4] Indeed, Quiccheberg praised the Fugger collections, which he had helped to build when he entered their service as librarian in 1555. He commented that their collections helped people think up new inventions.[5] Even before these merchant collections developed, it appears that the artisans of Augsburg and Nuremberg held extensive collections. Quiccheberg claimed that his own collection was already far surpassed by the collections of tools and precious items possessed by the artisans of these Free Imperial Cities: "It happened, as I truly confess, that I was energetically surpassed by goldsmiths, painters, sculptors, and other almost illiterates."[6]

Whatever the dates of the first secular collections—and we should not forget the collections of the dukes of Burgundy from the fifteenth century[7]—Quiccheberg's treatise has been viewed as marking an important moment in the history of collecting, for it represents the first written treatise, in contrast to the actual collections themselves, that suggests a systematic organization of a collection.[8] A close reading shows, however, that Quiccheberg's treatise is pragmatic rather than systematic, no doubt reflecting its composition in the service of the Wittelsbach duke, Albrecht V of Bavaria, and the fact that the wishes of this patron were uppermost in Quiccheberg's mind.[9]

Quiccheberg, who had been born in Antwerp, raised in Nuremberg, and educated in Basel and Ingolstadt, entered the service of Duke Albrecht V of Bavaria in 1559. In 1565, the same year that he wrote the treatise, the duke ordered seventeen precious objects to be kept in perpetuity by the Wittelsbach family. A princely proclamation of this sort was necessary in order to protect them from being melted down or sold as the need for specie arose. The explosive growth of such collections in the course of the century is illustrated by the fact that Albrecht's original list of items was increased to twenty-seven a few years later, and then still further by Duke Wilhelm V, finally reaching 3,407 items by 1598, only thirty-three years later.[10] Such collections were filled with (as the inventory of Rudolf II's *Kunstkammer* classified them) *naturalia*—things of nature; *artificialia*—things made by the human hand; and *scientifica* or *instrumenta*—instruments of all kinds.[11] In accord with different interests and regional concerns, patrons often specialized in the collecting of different objects. For example, Archduke Ferdinand's Schloss Ambras collection focused on armor, *naturalia* (especially the wonders of nature), and works of art that incorporated natural objects. The Saxon Elector, August I, was particularly interested in tools and objects associated with mining and metalworking; he accumulated an impressive collection of goldsmithing tools, including a wire-pulling machine and many other items, now held by the Musée de la Renaissance at Château d'Ecouen outside Paris. He also collected measuring and drawing implements and a remarkable collection of practical and scientific books, which he bought at the rate of about a hundred per year throughout his entire thirty-three-year reign.[12]

All these collections functioned not only as reservoirs of specie and demonstrations of noble largesse and living, but also as vehicles for self-definition and high politics. Collecting was about princely mastery, including the mastery of nature, and Quiccheberg's 1565 treatise, the

"Ample and all-encompassing Theater," begins in this manner, making clear that a collection should be deployed in the representation of the power, wealth, and noble characteristics of the ruler. Quiccheberg's treatise divided the collection into five "classes" of inscriptions—it must be remembered that Quiccheberg is not so much describing a space as describing a set of objects and their labels (*"Inscriptiones vel Tituli Theatri Amplissimi"*) that will be set into a space probably containing multiple chambers. Although he lists five classes of labels, and thus his scheme has been compared to Giulio Camillo's memory theater, he is not so much interested in the scheme of organization as in the material objects themselves. This emerges both in his comments on the classes and in the fact that he often refers to material aspects of the objects, for example, their size, in justifying their classification within the various classes.

In the first class Quiccheberg placed saints' portraits, since the saints were intermediaries between the rulers, representing both the divine on earth and God. Following these came family portraits, then maps that delineated the ruler's territory. Along with these could be included maps of the world, but Quiccheberg specified that the map of the patron's territory must be more conspicuous, larger, and more richly ornamented than all the rest. The political and representational aspects of these collections, as demonstrated by Quiccheberg's remark about the maps, is vitally important and much has been written about it, but more significant for this volume's focus on nature is that Quiccheberg does not dwell on political representation. Instead he moves rapidly over paintings of the ruler's residence, military campaigns, ritual, and spectacle to the things that appear closer to his heart: paintings of large, rare, or unusual animals, especially those found in the patron's territory, and all kinds of models demonstrating human artifice. These models, which were meant to be scaled up for the use of architects and builders, included buildings of all types, constructed of wood, paper, and feathers and ornamented with colors, as well as models of ships, wagons, stairs, fountains, arches, and bridges. In addition, small-scale models of machines, including water pumps, sawmills, grain mills, stamping mills, and dams, were to occupy this first section of the collection, expressly made for the purpose of scaling them up to full size and thereby discovering whether they functioned usefully or could be improved.[13]

The first section of the collection, which focuses on representation of all kinds, both of political power and of the workings of art (in the sense of *ars*, or the work of the human hand), is followed by four more sections, which all center on the relationship between nature and human *ars*. The second set of titles begins with the representation of power embodied by statues of emperors, kings, famous men, divine beings, and even animals, but appears more concerned with the material out of which the statues are formed, including stone, wood, white clay, marble, and all kinds of metals. Quiccheberg again passes quickly over these to the arts by which materials are worked, alluding to metalworking, turning, sculpting, woodworking, glassblowing, embroidery, weaving, tools and machines, and vessels of all sorts—foreign, ancient, and religious—as well as weights and measures and everything connected to agriculture and mining, not forgetting the staple objects of collections in this period: coins and portrait medals. Between a discussion of emblematic medals and copper engraving plates (in which he appears to have had a great interest), Quiccheberg alludes to small figures made by goldsmiths, including little ornamental sculptures with leaves, flowers, animals, and shells. Quiccheberg is here making reference to life-casts, such as those made by his contemporary and acquaintance, the Nuremberg master goldsmith Wenzel Jamnitzer, which were very much sought after by collectors (see fig. 1).

The third set of titles contains natural materials of all sorts, including preserved animals, raw materials of metalsmithing, seeds and herbs, colors and pigments to dye metal, resins, wax, sulfur, wood, ivory, textiles, and woods, as well as earthy materials including "juices of the earth," both natural and artificial, particularly medicinal earths, chalks, clays, vitriol, alum, salt, and fluids from dripping hot springs and grottos. Quiccheberg includes in this section prostheses for human limbs as well as parts of animals, such as horns, snouts, teeth, bones, bezoar, bladder and kidney stones, pelts, feathers, claws, skins, and skeletons. It appears that Quiccheberg's principle of organization in this grouping is materials and structures that are part of organic or transformative processes. This explains the presence in this class of impressive stones, such as marble, jasper, alabaster, and porphyry, for all these striated stones show evidence in their appearance of having grown in the earth. At the same time, Quiccheberg again includes in this section casts of

Figure 1. *Life-cast of a Lizard*, sixteenth century, lead, Staatliche Museen zu Berlin–Preussischer Kulturbesitz (Kunstgewerbemuseum), inv. no. K 5912. Photo by Jörg P. Anders. By permission.

animals made from metal, plaster, ceramics, and other naturalistic objects. Here, lifelikeness seems to be the goal, and he comments, "By this art, they all seem to be alive, for example, lizards, snakes, fish, frogs, crabs, insects, shells . . ." and they are colored so that "one believes they are real."[14] Life-cast plants and silk flowers are also included in this section. We shall return to these lifelike objects below.

In the fourth section of the collection, instruments, craftspeople's tools, and machines of all kinds were to be displayed, including instruments for writing, surveying, hunting, gardening, making war, making music, raising weights, undertaking surgery and dissection, and indeed all the tools by which "artisans all over the world in our time nourish the world."[15] In this section Quiccheberg also includes foreign clothing, which would be displayed on dolls, and almost as an afterthought, unusual or rare clothing, especially that belonging to the patron commissioning the collection.

The fifth and final section contains all types of images in different media, such as oil paintings, watercolors, tapestries, engravings, genealogies, portraits of famous men, coats of arms, and inscriptions, which were to be both painted on walls and hung on little panels. In this section, too, are listed the numerous cabinets in which small items could be stored and displayed.

Attached to this Universal Theater, Quiccheberg specified the addition of a library, which was to be organized according to the subject matter of the books;[16] a printing press; a lathe room, which already existed from the time of Duke Wilhelm IV (d. 1560);[17] and a medicine cabinet, which he noted was especially the province of women because they desired to help the sick and poor. Quiccheberg praised medicine as a constant delight for the mind because it always leads to new experience, and observed that the Wittelsbach court already possessed such a medicine cabinet, established through the work of Duchess Anna, the wife of Duke Albrecht. According to Quiccheberg, Duchess Anna also maintained a large aviary holding many different types of birds, and she allowed scholars access to it to study and gain knowledge about the birds.[18] Workshops, too, were to be attached to the collection, including a casting and stamping operation with both a smith's forge and an alchemical furnace, which had the particular use of producing life-casts of plants and animals. Such casts would preserve and stand in for these decomposable items.[19] It is not without significance that the alchemical furnace was to be used for life-casts, a point to which we shall return below.

Quiccheberg counsels the expansion of the collection by rooms that would respond to a patron's special interests, such as musical instruments, chests of clothes and ornaments for plays and masked balls, or weapons. At the same time, Quiccheberg assured the less flush collector that he did not have to include all sections of the collection—just those parts he could afford or that coincided with his interests. But, it was essential for a collector to employ a person he can send into all regions of the earth to seek out wonderful things, as well as individuals who are knowledgeable about the objects in the collection.[20] Quiccheberg also gave advice on how to organize the objects. He dismissed Pliny because Pliny wrote for philosophers, not for princes; he put aside Vitruvius's and Giulio Camillo's (1480–1544) organizing principle of the seven planets, and advised instead a "simpler" organization, one that adhered to the "form" of the thing (*formas rerum*), by which he seems to mean the object's general appearance, based both on its material and function.[21] Quiccheberg's principle of ordering, however, is far from rigorous.

From this brief overview, it is clear that Quiccheberg's ideal collection, while working to glorify and represent the mastery and power of the collector, focused primarily on the artifice of nature and of the human hand: on craftspeople, their tools, materials, and products. Quiccheberg attended to the material and the mechanic in his treatise. Is this an attempt to imitate the macrocosm, as some scholars have asserted? Let us now turn to the work of one of those "almost illiterates" to whom Quiccheberg referred, to gain insight into what the material and mechanic might have meant to Quiccheberg.

Quiccheberg mentions Wenzel Jamnitzer in the *Inscriptiones vel Tituli Theatri Amplissimi* as an artisan-collector.[22] Born in Vienna, Jamnitzer learned goldsmithing there under his father before he moved to Nuremberg and became both a member of the Nuremberg goldsmiths' association and citizen of Nuremberg in 1534. During his life, Jamnitzer played a leading role among Nuremberg goldsmiths and assisted the city council with information and regulation of trade.

In 1556 Jamnitzer began an ambitious "chamber fountain" for Emperor Maximilian II, but only delivered it twenty-two years later to Rudolf II in 1578. As in the case of Quiccheberg's theater, scholars have viewed this fountain as a representation of the cosmos, and it seems cosmic indeed. Ten feet high and five feet across, the fountain was assembled within a room (hence the term *Zimmerbrunnen*, or chamber fountain), and it comprised several different tiers, each pertaining to elements of the divine, human, and political cosmos. It possessed, according to a description probably written by Jamnitzer, "not only physics and metaphysics, but also politics, as well as many wonderful philosophical and poetical secrets displayed and proven to the eyes."[23] By examining this fountain as a collection, as a theater or storehouse analogous to Quiccheberg's theater of creation, we can begin to understand the aims of Quiccheberg's collecting and why he might have considered it a "first philosophy."

Jamnitzer's fountain, according to its description, sat on four figures representing the four seasons—Flora (spring) with her bouquet of flowers, Ceres (summer) holding a cornucopia and crowned with a wreath of ripe wheat, Bacchus (fall) holding a bunch of wine grapes, and Vulcan (winter) with a plow he has just hammered out on his anvil—which show, according to the description, the inalterable band of nature that forms a framework for all of human life.[24] These four statues were the only part of the fountain not melted down in the eighteenth century, a fate shared by nearly all table fountains and the vast majority of medieval and Renaissance objects made from precious metals. Above these figures, the fountain rose in five tiers, in accord with the structure of nature based on the four elements.

The lowest tier symbolized earth and was represented by Cybeles, goddess of the earth and a daughter of Saturn. She was surrounded by a grotto and all kinds of mining implements, including native silver and gold. From between the ores and the silver formations grew silver and gold flowers, cast from life, looking as if they were growing naturally along lively little brooks flowing with water between the rocks. Along the brooks, a polishing mill, stamp mill, sawmill, and hammer mill—all driven by the water—functioned in miniature. Here too were the little animals cast from life for which Jamnitzer was especially known and admired.

The next tier was a basin symbolizing the element of water, represented by Neptune standing on a shell drawn by hippopotamuses around the basin, and battling strange sea monsters that first moved toward him threateningly and then fled from him. The constant movement and the to-and-fro of the battle signified both the ebb and flow of the sea around the earth and the fact that great lords and potentates must battle constantly with enemies of the common good. This section was surrounded by a crown, reminding rulers that they must carefully keep secret their plans of war.

The third tier, symbolizing air, was represented by Mercury, who swung and swooped off the fountain as if actually in flight. Under him a dark cloud spewed raindrops, and images of the four winds portended a furious storm. Flying into the storm were all sorts of birds, symbolic again of air, and four angels carrying laurel wreaths, signifying the serving spirits that intervene between God and humans and reminding the viewer that God will protect and maintain his children in even the fiercest storm.

Above the tier of air was that of fire, represented by Jove. Having reached this height, the fountain takes on a complex admonitory political program which comprises a mirror for the prince. It is devoted to showing the order (*polizei*) preserved by the emperor and the hierarchy of nobles as the representatives of god on earth. Surrounding Jove were four angels arranged in such a way that, as Jupiter circles among them, each bows in reverence before they turn outward again to the human race, to whom they have been destined as serving spirits. The angels make clear how any high potentate should organize his regime: as a Michael, a strong hero who protects the subjects and the whole land; as a Gabriel, a wise, powerful speaker and chancellor; as a Uriel, a righteous judge and good preacher who brings the truth to light through his words; and finally as a Raphael, a learned and true physician and an experienced teacher of all the liberal arts. Four eagles fly between the angels, equipped with scepters, to signify the characteristics of the ruler, who among other traits is merciful and generous to "small birds, or subjects."

Above the angels and eagles stands a single large eagle, strangling a basilisk with one claw and holding a stone in the other. Under one wing it holds a scepter and in the other a small flag that moves up and down to honor God. On this eagle rides Jupiter, Lord of lords, who holds a lightning bolt in his right hand and pours out water from a vessel from his left. With these implements, he punishes evil with bolts of fire and blesses the good with the streams of water that bring fertility. On his head is a crown with eight points, the source of small streams that well up and gently sprinkle down in a rain of fertility and healing. But in a reflection of his majesty, a flame of fire shoots from his head.

Under this heavenly firmament rise four arches, symbolizing the Imperial crown, beneath which sit four monarchs, the last of which, the Holy Roman Emperor, holds a scepter and globe. Around these monarchs is a complete representation of the political structure of the Empire and the lineage of the Austrian house up to Rudolf II. The description of this magnificent fountain ends on a much less elevated note, with an account of the music for two peasant dances, the "Rolandt" and the "Pickelhäring," which begin to play when the water flow is set in motion.[25]

As a viewer beheld Jamnitzer's fountain and his eyes traveled down it, he would first take in the representation of the power and reach of the House of Habsburg, as well as the political structure of the Holy Roman Empire. He would consider the lessons of prudence that the fountain imparted to ruler and subject alike,[26] and he would absorb the description of nature as made up of four elements in constant ebb and flow. As he came to the bottom tier of earth, with its arresting gold and silver mines filled with ore and native metals, and with the life-casts of animals and plants arranged around flowing streams along which miniature mills operated, he might wonder about the processes of metamorphosis by which ore becomes gold and silver, or about the mechanics of water flow, or the workings of mechanical devices.[27] Below these, the fountain rested on the personifications of the seasons symbolizing the natural processes by which all things come into being and pass away. This fountain contained the same range of objects as in Quiccheberg's theater—natural materials, the works of the human hand, and objects of political representation—and both were intended to engage the viewer in active consideration, but Jamnitzer's fountain gives a surrounding context to the objects and materials that were displayed as separate elements in Quiccheberg's theater. More than anything else, the viewer of Jamnitzer's fountain and the visitor to Quiccheberg's theater would be confronted with the power of human art. Indeed, Jamnitzer's fountain embodied and displayed, in a single and unified work of art, the powers of nature and art to which Quiccheberg's collection alluded.

Jamnitzer's other works of art also embody and display the powers of art and nature. His Merckel Table Decoration of 1549 featured grasses and flowers, cast from life and springing from an egg-like vessel at the top, while around the base, reptiles, also cast from life, creep forth from the earth.[28] The central female figure represents mother earth, emblazoned with the words "I am the Earth, mother of all things, beladen with the precious burden of the fruits which are produced from myself,"[29] and the whole piece symbolizes the fertility and generative powers of nature. But, more than this, it demonstrated

Jamnitzer's own powers of creation and his ability to imitate nature. He made a similar demonstration in an extraordinary sculpture depicting the moment at which Daphne was turned into a laurel tree to escape the unwanted attentions of Apollo. Jamnitzer's silver and gilt Daphne rises out of tiny specimens of ore at its base; her arms, raised in despair, have been transformed into blood-red coral tree branches, just beginning to sprout tiny green enameled leaves. This is, of course, not merely a demonstration of Jamnitzer's ability to harness the processes of nature and to manipulate a variety of natural materials, but also a meditation on the processes of metamorphosis itself: the smelting of ores, the flow of metals and blood, the generation of new life.

The technique of casting plants and small animals from life was reinvented in Nuremberg in the sixteenth century, where Jamnitzer was its most active and well-known practitioner. A remarkable anonymous goldsmith's treatise, probably written in Toulouse and contemporaneous with Jamnitzer, contains detailed instructions for preparing the materials, such as the sand and plaster, for the mold, then catching the animals alive, keeping them, killing them, affixing them to the base of the mold, constructing the mold and the investment material, lifting the animal out of the box mold or burning out the creature, then casting the metal in the mold, and removing the sculpture from the mold. Generally, animals were killed by immersion in vinegar and urine so that they would not be deformed by blows, then posed in a lifelike manner by attaching them with pins and threads to a clay base. A thin plaster and sand solution was painted over them, and the whole thing was then fired in a kiln, which hardened the plaster and burned out the organic matter. This formed a mold that was first cleaned out with mercury or by blowing and then poured with metal. Dead animals might also be pressed into a sand mold, out of which they could be lifted before the molten metal was poured in.[30]

I have discussed elsewhere the significance of casting from life as a way of knowing nature. This style offered artisans the opportunity to display their art—their ability to imitate nature—both because the finished product was a perfect imitation of nature, and because they imitated the processes of nature in the smelting of the ores and the manipulating of metals in casting. This imitation of nature comprised a form of natural knowledge, both in the techniques used to produce it as well as in the epistemological claims made by the artisans.[31] But the anonymous goldsmith's manual makes the *investigative* nature of casting from life particularly clear. For example, it discusses the processes of catching lizards and snakes:

> Take a stick, pin a net with a slipknot to the top. Whistle and move the net nearer to the head of the lizard, and pull when it puts its head inside the net. It is more difficult to take a lizard with your hands than a snake, because lizards bite without letting go, with a bite as strong as pincers.
>
> You can take snakes with your hand, but cover your hand with a woolen cloth because the teeth of the snake can go through common cloth. You can recognize dangerous snakes by their blue eyes. They do not bite into water, as is known by crayfish catchers.[32]

And it treats the behavior of animals:

> TO MOLD SNAKES: Before molding your snake . . . do not remove its teeth, for [then] . . . snakes suffer gum pain and cannot eat. Keep your snake in a barrel full of bran, or, better, in a barrel full of earth in a cool place, or in a glass bottle. Give your snake some live frogs or other live animals, because snakes do not eat them dead. Also I've noticed that when snakes want to eat something or to bite, they do not strike straight on, on the contrary they attack sideways as do Satan and his henchmen. Snakes have small heads, but very large bodies, they can abstain from eating for 7 or 8 days, but they can swallow 3 or 4 frogs, one after the other. Snakes do not digest food all at once, but rather little by little. . . . If you worry and shake your snake, it will bring up digested and fresh food at the same time. Sometimes 2 or 3 hours after swallowing a frog, it can vomit it alive.[33]

Alongside such explicit natural historical observations are numerous experiments on the behavior of sands, clays, and firing techniques, as well as directives for the best methods of casting reptiles. For example, in discussing various kinds of sands to be used in box molds, the goldsmith comments, "The powder from the millstone of the edge tool maker is very good to cast with copper, but do not use the powder from the cutler."[34] At another point, he explicitly refers to the experimental nature of his work: "Since my last experiences, I molded with burned bone, clinker and burned felt."[35]

We can see, then, that this goldsmith sought out the behavior of animals and natural materials in a systematic and empirical way. The techniques by which casting from life was achieved involved significant investigation of this type. This is echoed in other artisans' manuals, which advise constant trial and experimentation.[36] In his works of art, Jamnitzer displayed not only the results of his natural investigation but also the products, indeed proofs, of his knowledge of the natural processes of metamorphosis and transformation.

When viewed against the background of Jamnitzer's works, Quiccheberg's third class of objects—an apparent jumble of natural materials, from "juices of the earth," parts of animals, bezoar stones, and striated stones, to life-casts of animals and silk flowers—begins to take shape as a similar project to Jamnitzer's. Quiccheberg's life-casts in this section are related to the other items because their making involves the same processes of nature that brought the natural materials into being, processes that involved, among other components, the earthy materials, vitriol, alum, salt, and the "juices of the earth." Like Jamnitzer's life-casts of animals and plants, these lifelike objects stood in for the real thing, but they also constituted a proof that natural processes are imitable and knowable by the human hand, and the objects themselves are a demonstration of the knowledge of these natural processes. We may recall that the life-casts of animals and plants were to be undertaken in the alchemical furnace, no doubt because fine gold and silver would be worked there, but also because alchemy was the science par excellence that imitated nature.[37] Examining the meaning of casting from life gives insight into the overarching aims of collecting nature and art in the mid-sixteenth century, aims that appeared to be shared by artisans and scholars.

The historian Dirk Jansen has pointed out that the *activity* of collecting was central for Quiccheberg, in contrast to Giulio Camillo, for whom ordering—a conceptual process—was the focus.[38] Quiccheberg's conception of a new "first philosophy" was active rather than being solely textual and conceptual, and, as we have seen in the comparison with Jamnitzer's work, it necessitated engaging with matter itself in order to come to know nature.

Quiccheberg's interest in crafts as the basis of a new "first philosophy" was not unique. Several scholars saw the crafts as a source for a new philosophy.[39] In the calls for intellectual reform of the sixteenth century, scholars attempted to found a new philosophy based on real things rather than words and gained by means of an active engagement with nature. Natural materials and natural objects—things to be apprehended by the five senses—became central. At about the same time that Quiccheberg was organizing the duke's collection and writing his treatise, the Parisian pedagogical reformer Peter Ramus (1515–72) visited artisans' workshops in Nuremberg for four days, insisting that he learned much there. He believed that he would be able to unify theory and practice through studying the combination of practical mathematics and artisanal production in Nuremberg.[40]

Quiccheberg, on the other hand, believed that he could bring the artisanal activity into the *Kunstkammer* and its associated workshops. Whether scholars visited workshops or urged princes to found workshops, the processes and techniques observed in these sites were difficult to capture in written texts. When the Nuremberg publisher Walther Ryff attempted to describe the technique of casting, he ended in frustration with the comment, "how much easier it is to understand from instruction on the spot than from written report."[41] Pomponius Gauricus, in his work on sculpture of 1503, the book on which Ryff based his treatise, had said that he would pass over the ugly, smoky part of bronze casting—or *Chemike*, as he called it in Greek—because it dealt with clay, coals, dung, and bellows. Besides, he was sure that his readers would be better off seeing the process than reading about it.[42] The inability of these scholars to set down in written form the work of artisans stemmed partly from the lack of a language to describe the necessary experience. Artisanal knowledge was tacit, no language that could convey experiential knowledge existed, and, while scholars were expert with words, texts, logic and disputation, they did not have much of an idea about how to approach the things and materials of nature, nor of how to go about doing and making. This disconnect between scholars and artisans reflected a particularly persistent feature of western culture that divided those who worked with their minds—scholars—from those who worked with their hands—artisans. This division was both intellectual and social, stretching back to the Greek disdain for manual work as deforming to mind and body.[43]

Scholars like Ramus and Quiccheberg persisted in the face of these difficulties and prejudices because they saw great potential in the productive knowledge of artisans. All manner of individuals—physicians, scholars, princes, and city governors—came to regard artisans as possessing the key that would unlock the productive powers of nature. The economic benefits of harnessing these powers were represented to princely patrons as a fundamental part of the prudence necessary for statecraft, and they believed that the collection of artisanal objects, tools, and workshops would reveal these secrets to them. Quiccheberg believed collecting was a first philosophy, a metaphysics that could open up the doors of wisdom in all the sciences; he noted that objects taught far more than books because memory was impressed more deeply by things and pictures than by books.[44] He also believed that the *Kunstkammer* would generate new inventions. A collection made according to Quiccheberg's treatise would thus display Nature as the source of productive knowledge and would disseminate the view that productive knowledge was the business of princes; that artisans held the key to natural knowledge; that handwork must be integrated into the system of knowledge; and, finally, that imitation was important as a means of transmitting knowledge and knowing nature. But more than this, the visual claim in naturalistic artworks that art imitated nature was a statement about the powers of both art and nature. Jamnitzer's sculptures also transmitted such claims about the possibility of knowing nature and the power of art that resulted from this knowledge.[45] His naturalistic works of art, like the small animals cast from life, expressed claims about his knowledge of nature and his mode of investigating it and made such connections between nature and art especially clear. Collections such as Quiccheberg's and works of art such as those created by Jamnitzer thus helped in a very direct way to form ideas about nature as productive and the potential of natural knowledge to harness that productivity. In the process of assembling *Kunstkammern*, then, scholars and their patron-princes absorbed and disseminated a new epistemology—a mode of gaining knowledge about nature—that originated with artisans and practitioners, and that eventually became central to the "new philosophy" or the new active science of the late sixteenth and seventeenth centuries.[46]

Quiccheberg, however, remained a scholar. His musings about what actually happened in the *Kunstkammer* and its associated workshops reveal him to be a typical word- and text-based scholar of his time, who could enumerate what artisans would learn from collections: "artisans of every type can practice distinguishing the difference between individual materials that they work,"[47] but could only articulate philological goals for scholars like himself. "I have recognized how pleasurable it is to visit individual craftsmen, to observe their admirable works and to research when German names can be compared and made equivalent to Latin names."[48] From such quintessentially scholarly remarks, I suggest that, despite their belief that they possessed a new philosophy—a new science that surpassed the learning of artisans—Quiccheberg and other sixteenth- and seventeenth-century scholars never harnessed this productive knowledge for themselves. Rather, scholars continued to rely on those "almost illiterates" until well into the nineteenth century.

Notes

1. Harriet Roth, ed., *Der Anfang der Museumslehre in Deutschland: Das Traktat "Inscriptiones vel Tituli Theatri Amplissimi" von Samuel Quiccheberg* (Berlin: Akademie Verlag, 2000), facing Latin and German text, p. 187.

2. From the work of Julius von Schlosser, *Die Kunst- und Wunderkammern der Spätrenaissance* (1908; repr., 2 vols., Braunschweig: Klinkhardt und Biermann, 1978), and Oliver Impey and Arthur MacGregor, eds., *The Origins of Museums: The Cabinet of Curiosities in Sixteenth- and Seventeenth-Century Europe* (Oxford: Clarendon Press, 1985), the study of early modern collecting has expanded exponentially. Notable recent contributions are Thomas DaCosta Kaufmann, *The Mastery of Nature: Aspects of Art, Science, and Humanism in the Renaissance* (Princeton: Princeton University Press, 1993); Horst Bredekamp, *The Lure of Antiquity and the Cult of the Machine*, trans. Allison Brown (Princeton: Markus Wiener, 1995); Paula Findlen, *Possessing Nature* (Berkeley and Los Angeles: University of California Press, 1994); Martin Kemp, *The Science of Art: Optical Themes in Western Art from Brunelleschi to Seurat* (New Haven and London: Yale University Press, 1990); and Martin Kemp, "'Wrought by No Artist's Hand': The Natural, the Artificial, the Exotic, and the Scientific in Some Artifacts from the Renaissance," in *Reframing the Renaissance: Visual Culture in Europe and Latin America 1450–1650*, ed. Claire Farago (New Haven and London: Yale University Press,

1995), 117–96; Ellinoor Bergvelt and Renée Kistemaker, eds., *De wereld binnen handbereik: Nederlandse kunst- en rariteitenverzamelingen, 1585–1735* (Zwolle: Waanders Uitgevers, 1992); and Andreas Grote, ed., *Macrocosmos in Microcosmo: Die Welt in der Stube. Zur Geschichte des Sammelns 1450 bis 1800* (Opladen: Leske & Budrich, 1994). See also Krzysztof Pomian, *Collectors and Curiosities: Paris and Venice, 1500–1800*, trans. Elizabeth Wiles-Portier (Cambridge: Polity Press, 1990), and Antoine Schnapper, *Le Géant, La Licorne et la Tulipe: Collections et Collectionneurs dans le France du XVIIe Siècle* (Paris: Flammarion, 1988).

3. Helen Watanabe-O'Kelly, *Court Culture in Dresden: From Renaissance to Baroque* (Hampshire and New York: Palgrave, 2002); Elisabeth Scheicher, *Die Kunst- und Wunderkammern der Habsburger* (Vienna: Molden, 1979); Elisabeth Scheicher, *Die Kunstkammer (Schloss Ambras)* (Innsbruck: Kunsthistorisches Museum, 1977); Christian Gries, "Erzherzog Ferdinand II von Tirol und die Sammlungen auf Schloß Ambras," *Frühneuzeit-Info* 5 (1994): 7–37.

4. See Mark A. Meadow, "Merchants and Marvels: Hans Jacob Fugger and the Origins of the Wunderkammer," in *Merchants and Marvels: Commerce, Science, and Art in Early Modern Europe*, ed. Pamela H. Smith and Paula Findlen (New York: Routledge, 2002), 182–200.

5. Roth, ed., *Der Anfang*, 193–95.

6. Roth, ed., *Der Anfang*, 125.

7. Jeffrey Chipps Smith, "The Artistic Patronage of Philip the Good, Duke of Burgundy (1419–1467)" (Ph.D. diss., Columbia University, 1979); Wim Blockmans and Walter Prevenier, *The Promised Lands: The Low Countries under Burgundian Rule, 1369–1530*, trans. Elizabeth Fackelman (Philadelphia: University of Pennsylvania Press, 1999); Dagmar Eichberger, *Leben mit Kunst: Wirken durch Kunst. Sammelwesen und Hofkunst unter Margarete von Österreich, Regentin der Niederlande* (Turnhout: Brepols, 2002).

8. It has been called the first museological text by Eva Schultz, "Notes on the History of Collecting and of Museums in the Light of Selected Literature of the Sixteenth to the Eighteenth Century," *Journal of History of Collections* 2 (1990): 205–18.

9. Quiccheberg includes a lengthy list of nobles, patricians, merchants, scholars, citizens, and artisans who collected (Roth, ed., *Der Anfang*, 164–211).

10. J.F. Hayward, *Virtuoso Goldsmiths and the Triumph of Mannerism, 1540–1620* (London: Sotheby Park Bernet Publications, 1976), 32.

11. Literature on the Habsburg and Rudolfine *Kunstkammern* is now abundant. For an overview and bibliography, see Eliska Fucikóva et al., eds., *Rudolf II and Prague: The Court and the City* (London: Thames and Hudson, 1997); Eliska Fuciková, ed., *Prag um 1600: Beiträge zur Kunst und Kultur am Hofe Rudolfs II* (Freren and Emsland: Luca, 1988); and the exhibition catalog, *Prag um 1600: Kunst und Kultur am Hofe Rudolfs II* (Ausstellung, Kulturstiftung Ruhr, Villa Essen, 10.6–30.10 1988).

12. See Watanabe-O'Kelly, *Court Culture in Dresden*.
13. Roth, ed., *Der Anfang*, 44–47.
14. Roth, ed., *Der Anfang*, 54–61.
15. Roth, ed., *Der Anfang*, 60–69.
16. Roth, ed., *Der Anfang*, 70–81.
17. Roth, ed., *Der Anfang*, printing press, 81; lathe room, 99.
18. Roth, ed., *Der Anfang*, 101–3.
19. Roth, ed., *Der Anfang*, workshops, 78 ff.; alchemical furnace, 83.
20. Roth, ed., *Der Anfang*, 93.
21. Roth, ed., *Der Anfang*, 111.
22. Roth, ed., *Der Anfang*, 205.
23. This marvel was melted down in the eighteenth century; only the Four Seasons that formed the base of this fountain still survive and are held by the Kunsthistorisches Museum in Vienna. See J. F. Hayward, "The Mannerist Goldsmiths: Wenzel Jamnitzer," *The Connoisseur* 164 (1976): 148–54, and Gerhard Bott, ed., *Wenzel Jamnitzer und die Nürnberger Goldschmiedekunst 1500–1700*, Catalog of the Germanisches Nationalmuseum, Nuremberg (Munich: Klinkhardt & Bierman, 1985). In the inventory of the *Kunstkammer* of Rudolf II, 1607–1611, the fountain is described as being contained in eighteen boxes marked with a particular seal. The contents of box no. 8 included a small book in which the entire meaning of the fountain was neatly written out on parchment. These documents are reprinted in Klaus Pechstein, "Der Goldschmied Wenzel Jamnitzer," in *Wenzel Jamnitzer und die Nürnberger Goldschmiedekunst 1500–1700*, ed. Bott, 67–70. A description of the fountain is contained in Hans Boesch, ed., "Urkunden und Auszüge aus dem Archiv und der Bibliothek des Germanischen Museums in Nürnberg," *Jahrbuch der kunsthistorischen Sammlungen der allerhoechsten Kaiserhauses* 7 (Wien, 1888), teil II, 87–90. See Miscellaneahandschrift Nr. 28722 in octavo, which contains sketches ("die auf Reisen in den Jahren 1640–42 gemacht wurden"), 87. The following description is from 87–88.
24. Boesch, ed., "Urkunden und Auszüge," 88.
25. Jamnitzer was known for his love of music. A Nuremberg composer dedicated a series of songs to him. See Sven Hauschke, "Wenzel Jamnitzer im Porträt: Der Künstler als Wissenschaftler," *Anzeiger des Germanischen Nationalmuseums* (2003): 127–36, esp. 136 n. 34.
26. A small casket-sized cabinet made by Jamnitzer, now in Madrid, also functioned as such a mirror for princes. The program of the ornaments on the cabinet makes clear the virtues a ruler must hold. See Viola Effmert, ". . . ein schön kunstlich silbre vergult truhelein . . . Wenzel Jamnitzers Prunkkassette in Madrid," *Anzeiger des Germanischen Nationalmuseums* (1989): 131–58.
27. Although the book of instructions for the fountain does not survive in an original copy, Jamnitzer's instrument manual, which is extant in the Victoria

and Albert Museum, London, as well as the manuals that accompanied other mechanical collections, explained the mechanical workings of the devices. For these other manuals, see Oscar Doering, "Des Augsburger Patriciers Philipp Hainhofer Beziehungen zum Herzog Philipp II von Pommern-Stettin: Correspondenzen aus den Jahren 1610–1619," in *Quellenschriften für Kunstgeschichte und Kunsttechnik des Mittelalters und der Neuzeit*, NF Bd. 6 (Vienna: Carl Graeser, 1894).

28. Klaus Pechstein, "Der Merkelsche Tafelaufsatz von Wenzel Jamnitzer," *Mitteilungen des Vereins für Geschichte der Stadt Nürnberg* 61 (1974): 90–121.

29. "Sum terra, Mater Omnium/Onusta caro pondere/Nascentium ex me fructuum." See the entire verse quoted in Pechstein, "Der Merkelsche Tafelaufsatz," 95–96.

30. Bibliothèque Nationale, Paris, Ms. Fr 640, R 62 039.

31. See Pamela H. Smith, *The Body of the Artisan: Art and Experience in the Scientific Revolution* (Chicago: University of Chicago Press, 2004).

32. Ms. Fr 640, R 62 039, f. 107r.

33. Ms. Fr 640, R 62 039, f. 109r.

34. Ms. Fr 640, R 62 039, f. 69v.

35. Ms. Fr 640, R 62 039, f. 86v.

36. See, for example, Vannoccio Biringuccio, *Pirotechnia* (1540), trans. Cyril Stanley Smith and Martha Teach Gnudi (Cambridge, Mass.: Harvard University Press, 1966), xvi.

37. As I discuss in *The Body of the Artisan*, chap. 4, alchemical practice was viewed as reproducing the processes of nature and alchemical theory as explaining transformation and metamorphosis.

38. Dirk Jacob Jansen, "Samuel Quicchebergs 'Inscriptiones': De encyclopedische verzameling als hulpmiddel voor de wetenschap," in *Verzamelen: Van rariteitenkabinet tot kunstmuseum*, ed. Ellinoor Bergvelt, Debora J. Meijers, and Mieke Rijnders (Heerlen: Open Universiteit; Houten: Gaade, 1993), 56–76.

39. Among the most well known of these scholars were Theophrastus Bombastus von Hohenheim, called Paracelsus; Juan Vives; and Peter Ramus. See Smith, *The Body of the Artisan*, chap. 2.

40. Reijer Hooykaas, *Humanisme, Science et Reforme, Pierre de la Ramee* (Leyden: E.J. Brill, 1958), 3–8 and 125. See also Paolo Rossi, *Philosophy, Technology, and the Arts in the Early Modern Era*, trans. Salvator Attanasio (New York: Harper and Row, 1970). For a wonderful example of the various kinds of knowledge sought in princely workshops, see Suzanne B. Butters, *The Triumph of Vulcan: Sculptors' Tools, Porphyry, and the Prince in Ducal Florence*, 2 vols. (Florence: Leo S. Olschki, 1996).

41. Gualterius Rivius, *De architectura* (Nuremberg, 1542), 95; quoted in Ernst Kris, "Der Stil 'Rustique': Die Verwendung des Naturabgusses bei Wenzel Jamnitzer und Bernard Palissy," *Jahrbuch der Kunsthistorischen Sammlungen in Wien*, NF 1 (1928): 141.

42. Pomponius Gauricus, *De Sculptura*, ca. 1503, trans. Heinrich Brockhaus (Leipzig: F.A. Brockhaus, 1886), 223.

43. See Smith, *The Body of the Artisan*, 7–8 and passim.

44. Roth, ed., *Der Anfang*, 139.

45. Smith, *The Body of the Artisan*, chaps. 2–3.

46. This is an argument that I have made, although without attention to Quiccheberg's treatise, in Smith, *The Body of the Artisan*.

47. Roth, ed., *Der Anfang*, 117.

48. Roth, ed., *Der Anfang*, 129.

CHAPTER SIX

"Procreate Like Trees"
Generation and Society in Thomas Browne's Religio Medici

Marjorie Swann

At one point in his meditative prose work *Religio Medici*, the Norfolk physician Thomas Browne takes on the role of sexologist—with deeply angst-ridden results. "I could be content," Browne says ruefully, "that we might procreate like trees, without conjunction, or that there were any way to perpetuate the world without this triviall and vulgar way of coition; It is the foolishest act a wise man commits in all his life, nor is there any thing that will more deject his coold imagination, when hee shall consider what an odde and unworthy piece of folly hee hath committed."[1]

For more than three centuries, this passage has scandalized and embarrassed many of Browne's readers. In 1643 the earliest commentator on *Religio Medici*, Sir Kenelm Digby, expressed incredulity that Browne yearned to "beget Children without the helpe of women or without any conjunction or commerce with that sweete, and bewitching *Sex*," and most later critics have likewise found Browne's sentiments "risible."[2] In his preface to *Religio Medici*, Browne carefully evades responsibility for any opinions his reader might find objectionable by cautioning, "There are many things delivered Rhetorically, many expressions therein meerely Tropicall, and as they best illustrate my intention; and therefore also there are many things to be taken in a soft and flexible sense, and not to be called unto the rigid test of reason" (60). Some commentators find in this disclaimer a license to ignore Browne's tree-envy: the sentiment *must* be meant as a joke.[3] In a similar vein, one scholar has recently argued that Browne's "bizarre" statements about sexual intercourse are best understood as the "juvenilities" of "a young man's book."[4] But Browne was not just a lad when he produced *Religio Medici*. The text was, it seems, written sometime between 1633 and 1635. Thus Browne, born in 1605, composed *Religio Medici* when he was in his late twenties, by which age he had graduated from Oxford, received his M.D. on the continent, and returned to England to serve his medical apprenticeship.[5] Moreover, Browne did not see fit to remove his praise of arboreal reproduction when he published the work about a decade later. Two unauthorized editions of *Religio Medici* appeared in print in 1642; in response, Browne published a slightly revised version of his text in 1643. Since first writing *Religio Medici*, Browne had married and become a father. Nonetheless, as he prepared *Religio Medici* for publication, Browne neither overlooked nor omitted his wistful comments about botanical procreation: aside from tweaking one verb, he let the controversial passage stand as originally written.

This biographical context perturbs many of Browne's critics and shapes their response to *Religio Medici*. Readers often feel compelled to create narratives—featuring Dorothy Mileham, who married Thomas Browne in 1641, and the couple's twelve children—to discredit the author's remarks about conjunction-free generation. One distinguished scholar divines that "[n]o slight was intended to [Browne's] young wife, just then expecting her first child, and we can be confident that none was taken," while another insists that we must not take Browne's ostensible distaste for copulation seriously, since "[t]his did not, in fact, affect the size of his family when he subsequently married, and it is clear that an extravagant literary style did not indicate an abnormal personality."[6] Another critic recently joined this chorus of biographical speculation by suggesting that Browne's minor revision of the passage—changing "I *wish*" to "I *could be content* that we might procreate like trees"—indicates Browne's increased enthusiasm for sexual intercourse:

"Does it mean that coition is a great pleasure to him (and to Dorothy Mileham), but he thinks they could live without it if required?"[7] Such comments usefully reveal the extent to which the critical response to *Religio Medici* has been informed—and limited—by readers' unspoken assumptions about human sexuality. Although the early modern church insisted "that all sexuality should be heterosexual, genital, and confined to marriage,"[8] we should consider that Browne's ideas may not be readily aligned with such norms.

In twitting Browne's distaste for copulation, one author wrote in 1691, "I wonder at the unnatural Phancy of such as could wish we might procreate like Trees, as if they were *asham'd of the Act*." Misguided men like Browne, the writer continues, "Had they been present with God, when he commanded *Adam* and *Eve* to encrease and multiply, they would have propos'd a better method for Generation."[9] But indeed, Browne and his contemporaries believed that God the Creator had devised more than one "method for Generation," establishing nonsexual processes of biological origin as well as the sexual "conjunction" by which Adam and Eve begat humanity. In early modern usage, the term "generation" thus not only refers to sexual reproduction, but more generally describes "the coming into existence of new individual organisms, both animal and plant, irrespective of the method which might be involved."[10] And throughout *Religio Medici*, Browne displays great interest in nonsexual modes of generation.

"I could be content that we might procreate like trees": rather than trying to write off this provocative statement, I propose, by contrast, to interpret *Religio Medici* in light of it.[11] I argue that generation "without conjunction" is central to the natural philosophy that permeates *Religio Medici*. Moreover, because concepts of the organization of natural processes are necessarily political, an understanding of Browne's depiction of generation must also lead us to reassess the politics of *Religio Medici*. Much recent analysis of Browne, following the lead of Michael Wilding, has found the writer's work informed by a conservative political stance.[12] But since the first publication of *Religio Medici*, readers have been unsettled by Browne's idealization of botanical generation precisely because it is so subversive of traditional social hierarchies. By taking seriously Browne's desire to procreate like trees, then, we may begin to appreciate how seventeenth-century theories of nonsexual generation could give rise to a potentially radical social vision.

For Thomas Browne, the natural world was filled with organisms that generated "without conjunction." In *Religio Medici,* Browne asserts that "*Natura nihil agit frustra*,[13] is the onely indisputable axiome in Philosophy; there are no *Grotesques* in nature; nor any thing framed to fill up empty cantons, and unnecessary spaces; in the most imperfect creatures, and such as were not preserved in the Arke, but having their seeds and principles in the wombe of nature, are every-where where the power of the Sun is; in these is the wisedome of his hand discovered" (77). This passage demonstrates Browne's belief in spontaneous generation. Since ancient times, natural philosophers as well as the uneducated accepted the notion that some animals developed, without parents, from mud, warm water, dew, or putrefying matter. "The generation of Animals is as various almost as their several Species," observed Nathaniel Highmore in 1651. Although higher forms of animals "owe their beginnings to some seminal parts derived from the Genitors," Highmore noted, others "derive their Pedegree from the corruption of Dirt, Mud, and other Animals; some arise from the funerals of Plants."[14] William Harvey concurred that "some animals are born spontaneously or, as is commonly said, out of putrefaction."[15] Frogs, fish, eels, and insects were widely understood to arise from such spontaneous (or "equivocal") generation. In linking spontaneous generation to the story of Noah's Ark, Browne follows St. Augustine, for whom the concept of spontaneous generation was exegetically useful, since it helped to explain how "the tiniest of creatures" were saved during the Flood: "it was not necessary for there to be in the Ark those creatures which can be generated from certain things, or from the corruption of such things, without sexual intercourse."[16]

For educated men of Browne's era, the natural world was thus populated by an abundance of parent-free creatures, "the first rank of Animals arising from corruption of other creatures (as Eeles from Mud; Flies and Wormes, from Beasts; the Scarabeus [a type of beetle] from Oxen; Lice from the filth of most Creatures)."[17] Indeed, the nonsexual fruitfulness of the natural world posed special health hazards for anyone foolhardy enough to drink stagnant water: "I have seen," reported Alexander Ross in 1651, "one whose belly by drinking of puddle water was swelled to a vast capacity, being full of small Toads, Frogs, Evets [newts], and such vermin usually bred in putrified water."[18] In Browne's opinion, however, spontaneous generation occurred only among lower animals. Hence, as his remarks about the reproduction of trees indicate, Browne did not share Paracelsus's belief that an artificial man, a "Homunculus," could be generated through the putrefaction of human sperm in horse dung; indeed, he explicitly refutes this idea in *Religio Medici,* saying, "I am not of *Paracelsus* minde that boldly delivers a receipt to make a man without conjunction" (106).[19] But, as he explains at length in *The Garden of Cyrus* (1658), not only do butterflies, moths, and flies generate spontaneously from plants, but, he attests, we should also acknowledge "the generation of Bees out of the bodies of dead Heifers" and "the production of Eeles in the backs of living Cods and Perches." For Browne, then, "the Æquivocall production of things under undiscerned principles, makes a large part of generation" (351).

It was the generation of plants, however, that especially captured Browne's imagination. Today we take for granted that plants reproduce sexually. We understand how the pistil in the center of a flower corresponds to the female sexual organs, how the pollen-laden stamens function as male reproductive structures, and how seeds are formed as fertilized ovules.[20] But not until 1676, when Nehemiah Grew delivered a series of lectures on botanical reproduction to the Royal Society, did Englishmen begin to realize that flowers, previously viewed as exemplars of innocence, had sex lives.[21] Writing before this sexualization of botany, Browne shares the orthodox view of plants as sexless entities. His notebooks reveal that he puzzled over the biological structure and purpose of flowers: "Noe man hath yet defined the true use of a flower, whether to protect the rudiment of the fruit or not, & fewer have determined the use of those efflorenceis & seeming seminallities wh. are observable in many vegetables, the same in maples, oakes, & the same in tulips, in roses, the blooms of apples; wh. cary litle bodyes on them wh. in magnifying glasses does lively represent seeds."[22] In *Pseudodoxia Epidemica* (1646), Browne concludes that plants, "wherein there is no distinction of sex," "multiply within themselves, according to the law of the Creation, Let the earth bring forth grasse, the herbe yeelding seed, and the tree yeelding fruit, whose seed is in it selfe. Which is indeed the naturall way of plants, who having no distinction of sex, and the power of the species contained in every *individuum,* beget and propagate themselves without commixtion."[23]

How did plants nonsexually "multiply within themselves," according to early modern English botanical theory? Browne and his contemporaries believed that some plants generated spontaneously. Robert Sharrock, for example, argued that in the prelapsarian world, all plants appeared "spontaneously without seed"; but after the Fall, as a result of "that Curse of God which man drew upon the Earth," beautiful and beneficial plants became more difficult to propagate, although "Thorns, Briars, Thistles, and other rude, imperfect, and almost useless Plants" continued to arise spontaneously.[24] Although Browne shares this belief in botanical "spontaneous productions" (351), he refers infrequently to the equivocal generation of plants. Instead, Browne is fascinated by the germination of seeds. Nathaniel Highmore theorized that plants imbibe from the earth a seminal "juice" that recreates the form of the plant in its seeds, a process Sir Kenelm Digby described in chemical terms:

> The juyce which was first in the button, (that is now become the fruit) and had passed from the root through the manifold varieties of the divers parts of the Plant, and had suffered much concoction and depuration, partly from the Sun and partly from the inward heat imprisoned in that harder stony part about the middle of the fruit; is by these passages, strainings, concoctions, and sublimations, become at length to be of the nature of a tincture extracted out of the whole plant; and is at last dryed up into a kind of Magistery, full of Fire and of Salt. This is that which we call the Seed; which being buryed in the Earth, and soaked with fitting humidity, in such sort as we have here at large declared, setteth on foot this work anew, and repeateth over again.[25]

For Browne, by understanding such botanical generation by seed, one can apprehend the divinely ordained principles that structure the natural world.

From Browne's perspective, part of the interest of seeds lies simply in their physical minuteness: "How little is required unto effectual generation, and in what diminutives the plastick principle lodgeth, is exemplified in seeds. . . . The exiguity and smallnesse of some seeds extending to large productions is one of the magnalities of nature, somewhat illustrating the work of the Creation, and vast production from nothing" (347, 351). Browne thus relives God's creation of the universe by studying the germination of a tiny seed. But if seeds are, for Browne, wonderful examples of *multum in parvo*, they are equally significant for their sheer ubiquitousness: "Rayne water wh. containeth seminall atomes elevated by exhalations making the earth fruitfull where it falleth."[26] Browne finds an image of perfection in "the dispersed Seminalities of Vegetables at the Creation scattered through the whole Mass of the Earth, no place producing all and almost all some" (430). Beneath the earth's surface, minerals are generated like plants, "determined by seminalities; that is created, and defined seeds committed unto the earth from the beginning."[27] Some seventeenth-century natural philosophers argued that spontaneous generation was very different from the development of seeds;[28] Browne, by contrast, often blurs the boundaries between the two modes of generation. As we have seen, in *Religio Medici*, Browne describes equivocal generation as proceeding from the effect of the sun on "seeds and principles in the wombe of nature" (77), and in *Pseudodoxia Epidemica*, Browne similarly depicts God as a kind of divine Johnny Appleseed, planting the germs of organisms in specially chosen areas of the global seed-plot: "For the hand of God that first created the earth, hath with variety disposed the principles of all things, wisely contriving them in their proper seminaries, and where they best maintaine the intention of their species; whereof if they have not a concurrence, and be not lodged in a convenient matrix, they are not excited by the efficacie of the Sunne."[29] These evocations of the sun catalyzing the development of seminal "principles" parallel accounts of the spontaneous generation of invertebrates from mud or carrion. However, Browne distinguishes "generations equivocall" from "seminall production" in terms of the relationship between the new organism and its origin: a creature produced from "corruptive" or spontaneous generation will not resemble its source (a maggot generated from a frog, for example), whereas an entity produced by "seminall power" will embody "the Idea" of its originator, as when plants come true from seed.[30] In *Pseudodoxia Epidemica*, Browne vehemently reiterates this principle, arguing that one must not take "putrifactive generations" as "correspondent unto seminall productions": "So when the Oxe corrupteth into Bees, or the Horse into hornets, they come not forth in the image of their originalls. So the corrupt and excrementous humors in man are animated into lyce; and we may observe that hogs, sheep, goats, hawkes, hens, and others, have one peculiar and proper kind of vermine, not resembling themselves according to seminall conditions, yet carrying a setled and confined habitude unto their corruptive originalls."[31]

This distinction between the mutability of identity characteristic of spontaneous generation and the transference of stable form through "seminall production" is important to Browne's thought as a whole. In *Religio Medici*, seeds exemplify unchanging identity: "In the seed of a Plant to the eyes of God, and to the understanding of man, there exists, though in an invisible way, the perfect leaves, flowers, and fruit thereof: (for things that are in *posse* to the sense, are actually existent to the understanding.) Thus God beholds all things, who contemplates as fully his workes in their Epitome, as in their full volume, and beheld as amply the whole world in that little compendium of the sixth day, as in the scattered and dilated pieces of those five before" (124). Browne is also captivated by the immutable endurance of seeds:

> To manifest how lasting the seminall principles of bodyes are, how long they will lye uncorrupted in the earth, or how the earth that hath been once impregnated therewith may retaine the powers thereof unto opportunity of actuation or visible production. A remarkable garden where many plants had been, being digged up & turned a fruitlesse ground, after ten yeeres being digged up, many of the plants returned wh. had layne obscure; . . . some plants will maintaine their seminallity out of the earth, as wee have tried in one of the least of seeds.[32]

In *Religio Medici*, Browne similarly contemplates the unalterable identity of plants through an exploration of palingenesis: "A plant or vegetable consumed to ashes, to a contemplative and schoole Philosopher seemes utterly destroyed, and the forme to have taken his leave for ever: But to a sensible Artist the formes are not perished, but withdrawne into their incombustible part, where they lie secure from the action of that devouring element. This is made good by experience, which can from the ashes of a plant revive the plant, and from its cinders recall it into its stalk and leaves again" (121). Infused with the same seminal juice that replicates its form in its seeds, an incinerated plant, Kenelm Digby asserted, "will admit no change into another Nature; but will always be full of the qualities and vertues of the Plant it is derived from." "If all the Essentiall parts could be preserved, in the severing and purifying of them, I see no reason but at the reunion of them, the entire Plant might appear in its complete perfection."[33] For Browne, plants thus exemplify the immutable quality of "Ideas," defined by another seventeenth-century writer as "the first Images of things, as they rise up from the Fountain of eternity. . . . The *Idea*, in this sense, is the first and distinct Image of each form of things in the Divine Mind."[34]

Within the framework of Browne's theory of generation, it is thus the nonsexual production of plants by seed, not postlapsarian human "conjunction," that most closely resembles God's creation of Adam and Eve. A passage from *Pseudodoxia Epidemica* deserves quotation in full, for it clearly demonstrates how Browne associates God's ability to create life nonsexually with the germination of seeds:

> There was therefore never any Autochthon, or man arising from the earth but Adam, for the woman being formed out of the rib, was once removed from earth, and framed from that element under incarnation. And so although her production were not by copulation, yet was it in a manner seminall: For if in every part from whence the seed doth flow, there be contained the Idea of the whole, there was a seminality and contracted Adam in the rib, which by the information of a soule, was individuated into Eve. And therefore this conceit applyed unto the originall of man, and the beginning of the world, is more justly appropriable unto its end; for then indeed men shall rise out of the earth, the graves shall shoot up their concealed seeds, and in that great Autumne men shall spring up, and awake from their Chaos againe.[35]

As Egon Stephen Merton has observed, "Browne's Idea, the potential individual, is immutable and eternal, the type or 'Exemplar' of which the actual individual is a copy. . . . Browne's Idea is both a potential individual and a Platonic reality."[36] And for Browne, the nonsexual generation of plants most nearly replicates the process by which God's Idea of human beings first became flesh. By contrast, the postlapsarian, sexual reproduction of humanity resembles spontaneous generation insofar as it transmutes Adam's offspring into lesser creatures: "Could we intimately apprehend the Ideated Man, and as he stood in the intellect of God upon the first exertion by Creation, we might more narrowly comprehend our present Degeneration, and how widely we are fallen from the pure Exemplar and Idea of our Nature: for after this corruptive Elongation from a primitive and pure Creation, we are almost lost in Degeneration; and *Adam* hath not only fallen from his Creator, but we ourselves from *Adam*, our Tycho and primary Generator" (430–31). With this understanding of the value Browne ascribes to botanical generation, the ability of plants to "procreate . . . without conjunction," we are ready to assess Browne's depiction of human relationships in *Religio Medici*.

Throughout *Religio Medici*, we find a tension between processes of generation that preserve identity and those that cause mutation. Browne seeks to resist forces that threaten to transform the self. In *Religio Medici*, Browne is thus content to understand himself as a glimmer in the eternal mind of God, forever guaranteed an immutable existence as an Idea: "I was not onely before my selfe, but *Adam*, that is, in the Idea of God, and the decree of that Synod held from all Eternity. And in this sense, I say, the world was before the Creation, and at an end before it had a beginning; and thus was I dead before I was alive, though my grave be *England*, my dying place was Paradise, and *Eve* miscarried of mee before she conceiv'd of *Cain*" (132). Here, Browne plays with the theory of preformation, the concept that "all the living things there were to be, had in fact been organized by God at creation, and that encapsulated within the first parent all future generations were present. The first female of every species, therefore, contained within herself all future generations of her kind."[37] For Browne, God the Creator is the source and guarantor of his identity. In a brief but provocative analysis of gender in *Religio Medici*, Daniela Havenstein observes that despite the importance of biological ideas throughout the treatise, Browne refers to women with striking infrequency, and "the notion of woman in *Religio Medici* is, moreover, limited to her procreative function."[38] What is particularly noteworthy about this passage, however, is the way it invokes Eve as Browne's biological mother only to deny the relationship: spontaneously aborted by Eve, Browne remains solely generated by God. Like Adam, God thus creates Browne as "the man without a Navell" (152).

In contrast to the immutable identity enjoyed by the miscarried Browne, the family man in *Religio Medici* endures an unstable existence. Browne depicts family relationships as fraught with the loss of affection, since conjugal "conjunction" necessarily generates emotional distance as well as biological offspring:

> Let us call to assize the lives of our parents, the affection of our wives and children, and they are all dumbe showes, and dreames, without reality, truth, or constancy; for first there is a strong bond of affection betweene us and our parents, yet how easily dissolved? We betake our selves to a woman, forgetting our mothers in a wife, and the wombe that bare us in that that shall beare our image. This woman blessing us with children, our affections leaves the levell it held before, and sinkes from our bed unto our issue and picture of posterity, where affection holds no steady mansion. They growing up in yeares desire our ends, or applying themselves to a woman, take a lawfull way to love another better than our selves. Thus I perceive a man may bee buried alive, and behold his grave in his owne issue. (160)

Given Browne's bleak view of family relationships in *Religio Medici*, his later evocation of a world populated solely by spontaneous generation takes on the quality of wishful thinking: "The probleme might have beene spared, Why wee love not our Lice as well as our Children, Noahs Arke had beene needlesse, the graves of animals would be the fruitfullest wombs; for death would not destroy, but empeople the world againe."[39] Browne also perceives the serial identity-shifting endemic to families in terms of spontaneous generation. In describing a dying patient in *A Letter to a Friend*, Browne recalls how the man began "to lose his own Face and look like some of his near Relations; for he maintained not his proper Countenance, but looked like his Uncle, the Lines of whose Face lay deep and invisible in his healthful Visage before: for as from our beginning we run through variety of Looks, before we come to consistent and settled Faces; so before our End, by sick and languishing Alterations, we put on new Visages: and in our

Retreat to Earth, may fall upon such Looks which from community of seminal Originals were before latent in us" (392). Here, Browne portrays family-ridden identity as an organism caught in the downward spiral of spontaneous generation, mutating ever further from its original state as it decays.

In contrast to the corruptive quality of family relationships, Browne finds in friendship the basis of stable identity. At times, *Religio Medici* becomes a hymn of praise to the sustaining power of friendship:

> There are wonders in true affection, it is a body of *Ænigmaes*, mysteries and riddles, wherein two so become one, as they both become two; I love my friend before my selfe, and yet me thinkes I do not love him enough; some few months hence my multiplyed affection will make me beleeve I have not loved him at all, when I am from him, I am dead till I bee with him, when I am with him, I am not satisfied, but would still be nearer him: united soules are not satisfied with embraces, but desire to be truely each other, which being impossible, their desires are infinite, and must proceed without a possibility of satisfaction. Another misery there is in affection, that whom we truely love like our owne, wee forget their lookes, nor can our memory retaine the Idea of their faces; and it is no wonder, for they are our selves, and our affections makes their lookes our owne. (143)

Rather than causing change, friendship reinforces identity; the beloved friend loses alterity, his individuated Idea disappears, and he instead merges with the desiring self. Browne admits that he loves his friends before his family, and more than he can imagine himself ever loving a woman:

> I confesse I doe not observe that order that the Schooles ordaine our affections, to love our Parents, Wifes, Children, and then our Friends, for excepting the injunctions of Religion, I doe not find in my selfe such a necessary and indissoluble Sympathy to all those of my bloud. I hope I do not breake the fifth Commandement, if I conceive I may love my friend before the nearest of my bloud, even those to whom I owe the principles of life; I never yet cast a true affection on a Woman, but I have loved my Friend as I

do vertue, my soule, my God. From hence me thinkes I doe conceive how God loves man, what happinesse there is in the love of God. (142–43)

And through friendship, the scholarly Browne can emulate his seed-planting God, generating in the minds of other men the contents of his own intellect: "I make not therefore my head a grave, but a treasure of knowledge; I intend no Monopoly, but a Community in learning; I study not for my owne sake onely, but for theirs that study not for themselves. I envy no man that knowes more than my selfe, but pity them that know lesse. I instruct no man as an exercise of my knowledge, or with an intent rather to nourish and keepe it alive in mine owne head, than beget and propagate it in his" (137–38).[40] In friendship, not the family, Browne finds the stable existence God bequeathed to plants—an identity which remains immutable because it can generate "without conjunction."

At the time that he wrote *Religio Medici,* Browne's distaste for procreation within marriage was ideologically subversive. During the early modern period, the conjugal family was viewed as the foundation of the social order.[41] Thus to escape from the normative demands of "conjunction" was to enter a world of transgression, a realm where the individual violated the most basic tenets of his society. In 1645, James Howell, although himself unmarried, criticized Browne for disdaining "the honourable degree of *marriage,* which I hold to be the prime Link of human society," echoing Kenelm Digby's criticism that Browne "setteth marryage at too low a rate, which is assuredly the highest and devinest linke of humane society."[42] John Evelyn, who admitted that he envied the ability of trees to "generate their like . . . without violation of virginity," nonetheless saw Browne's rejection of conjugal marriage as leading to a wholesale rejection of all social hierarchy: "Indeed if all the world inhabited the *Desarts,* and could propagate like *Plants* without a fair Companion; had we goods in common, and the *primitive* fervour of those new made *Proselites*; were we to be governed by *instinct*; in a word, were all the *Universe* one ample *Convent,* we might all be contented, and all be happy; but this is an *Idea* no where existant on this side of *Heaven*; and the *Hand* may as well say, . . . *I have no need of the Eye,* as the World be governed without these necessary subordinations."[43] In a communistic utopia, perhaps, Browne could "propagate" like the plants he so admired; but in this fallen world, procreative sex within marriage constitutes one of the "necessary subordinations" upon which society is founded. The unease which his botanical fantasy continues to elicit from readers of *Religio Medici* suggests that even in the twenty-first century, we are still uncomfortable with the radical potential of Thomas Browne's vision that "we might procreate like trees, without conjunction."

Notes

1. Thomas Browne, *Religio Medici*, in Browne, *The Major Works*, ed. C.A. Patrides (Harmondsworth: Penguin, 1977), 148–49. Unless otherwise noted, all future references to the works of Browne are to this edition, and page numbers are included, in parentheses, in the text. I have found James Eason's online, searchable texts of Browne's works very helpful (http://penelope.uchicago.edu). Since the title was granted decades after Browne wrote the works I consider here, I follow Claire Preston's example by not referring to Browne as "Sir" (*Thomas Browne and the Writing of Early Modern Science* [Cambridge: Cambridge University Press, 2005], 4).

2. Sir Kenelm Digby, *Observations upon "Religio Medici"* (London, 1643), 110–11; Raymond B. Waddington, "The Two Tables in *Religio Medici*," in *Approaches to Sir Thomas Browne: The Ann Arbor Tercentenary Lectures and Essays*, ed. C.A. Patrides (Columbia: University of Missouri Press, 1982), 97.

3. C.A. Patrides argues that "the lighthearted tone of the passage suggests that Browne again deploys his 'soft and flexible sense'" (Browne, *The Major Works*, 149 n. 77).

4. Preston, *Thomas Browne and the Writing of Early Modern Science*, 18.

5. On Browne's education, early career, and composition of *Religio Medici*, see Frank Livingstone Huntley, *Sir Thomas Browne: A Biographical and Critical Study* (Ann Arbor: University of Michigan Press, 1962), 1–71, 90–98.

6. Joan Bennett, *Sir Thomas Browne: "A Man of Achievement in Literature"* (Cambridge: Cambridge University Press, 1962), 9; Geoffrey Keynes, "Introduction," in *Religio Medici and Christian Morals* by Sir Thomas Browne, ed. Keynes (New York: Thomas Nelson, 1940), x.

7. Italics in quotations from Browne mine; Ronald Huebert, "The Private Opinions of Sir Thomas Browne," *Studies in English Literature, 1500–1900* 45 (2005): 123.

8. Patricia Crawford, "Sexual Knowledge in England, 1500–1750," in *Sexual Knowledge, Sexual Science: The History of Attitudes to Sexuality*, ed. Roy Porter and Mikuláš Teich (Cambridge: Cambridge University Press, 1994), 83.

9. Benjamin Bridgwater and John Dunton, *Religio Bibliopolae: In Imitation of Dr. Browns Religio Medici With a Supplement to It* (London, 1691), 77–78. Some scholars ascribe this work solely to John Dunton; I accept Daniela Havenstein's arguments for joint authorship (*Democratizing Sir Thomas Browne: "Religio Medici" and Its Imitations* [Oxford: Oxford University Press, 1999], 47).

10. Elizabeth B. Gasking, *Investigations into Generation, 1651–1828* (Baltimore: Johns Hopkins Press, 1967), 7.

11. Havenstein notes that "[t]here are at least three possible ways of dealing with this passage. Either we discard it as the 'unnatural Phancy' of a frigid young man, or we treat it as a joke, or we try to put in the broader context of *Religio Medici*" (*Democratizing Sir Thomas Browne*, 63); like Havenstein, I am taking the third approach. Scholars who assess seriously Browne's remarks about botanical reproduction include Michael Stanford, who argues that Browne regards the physical body as shameful ("The Terrible Thresholds: Sir Thomas Browne on Sex and Death," *English Literary Renaissance* 18 [1988]: 413–23), and John Rogers, who finds that Browne typifies his era's "scientific, vitalist fantasy of non-contactual agency" (*The Matter of Revolution: Science, Poetry, and Politics in the Age of Milton* [Ithaca, N.Y.: Cornell University Press, 1996], 93). By "vitalist," Rogers refers to "animist materialism" or the idea that "all material substance" has "the power of reason and self-motion" (*Matter of Revolution*, 1).

12. Michael Wilding, "*Religio Medici* in the English Revolution," in his *Dragon's Teeth: Literature in the English Revolution* (Oxford: Oxford University Press, 1987), 89–113. My analysis of Browne's natural philosophy complements my earlier argument that Browne's antiquarianism reveals the author's equivocal stance toward traditional forms of social authority (Marjorie Swann, *Curiosities and Texts: The Culture of Collecting in Early Modern England* [Philadelphia: University of Pennsylvania Press, 2001], 121–34).

13. "Nature does nothing in vain"; Browne here cites Aristotle (77 n. 87).

14. Nathaniel Highmore, *The History of Generation* (London, 1651), 57.

15. William Harvey, *Disputations Touching the Generation of Animals* (1651), trans. Gweneth Whitteridge (Oxford: Blackwell, 1981), 150. Elsewhere in this treatise, Harvey appears to reject the idea of spontaneous generation; but, as Elizabeth Gasking observes, Harvey's work as a whole indicates "an implicit acceptance" of spontaneous generation (*Investigations into Generation*, 19).

16. Augustine, *The City of God against the Pagans*, ed. and trans. R. W. Dyson (Cambridge: Cambridge University Press, 1998), 691. Other church fathers made different theological uses of spontaneous generation; Lactantius, for example, argued that the existence of such nonsexual reproduction gave credence to the doctrine of the Virgin Birth (Aram Vartanian, "Spontaneous Generation," in *Dictionary of the History of Ideas*, ed. Philip P. Wiener [New York: Charles Scribner's Sons, 1973], 4:308).

17. Highmore, *History of Generation*, 58. As Preston notes, "The advent of powerful microscopes after 1660 showing animalcules where none had been suspected tended to quash the theory of spontaneous generation, but even Hooke persisted with it although he had made microscopic observations to disprove it" (*Thomas Browne and the Writing of Early Modern Science*, 193 n. 63).

18. Alexander Ross, *Arcana Microcosmi* (London, 1651), II.x.5, p. 224. Browne himself remarked that rain water, "which appearing pure and empty, is full of seminall principles, and carrieth vitall atoms of plants and animals in it, which have not perished in the great circulation of nature, as may be discovered from severall insects generated in raine water" (Thomas Browne, *Pseudodoxia Epidemica*, ed. Robin Robbins [Oxford: Oxford University Press, 1981], 1:249).

19. On Paracelsus, see F.J. Cole, *Early Theories of Sexual Generation* (Oxford: Oxford University Press, 1930), 1–2. While castigating Browne for wishing "to propagate the world without conjunction with women," James Howell notes that "*Paracelsus* undertakes to shew him the way" (*Epistolae Ho-Elianae* [London, 1645], 91). In *Pseudodoxia Epidemica*, Browne similarly expresses skepticism about Johannes Baptista van Helmont's claim that mice can be spontaneously generated from grain fermented in the presence of dirty shirts (*Pseudodoxia*, 1: 290); on van Helmont's spontaneously generated mice, see Bentley Glass, "The Germination of the Idea of Biological Species," in *Forerunners of Darwin: 1745–1859*, ed. Glass et al. (Baltimore: Johns Hopkins Press, 1959), 39–40.

20. For a lucid, well-illustrated discussion of plant reproduction, see Brian Capon, *Botany for Gardeners*, rev. ed. (Portland, Ore.: Timber Press, 2005), 178–98.

21. Although the ancient Babylonians and Assyrians understood that date palms reproduce sexually, this knowledge did not lead to a broader understanding of plant sexuality (Conway Zirkle, *The Beginnings of Plant Hybridization* [Philadelphia: University of Pennsylvania Press, 1935], 3–7, 89–90).

22. Geoffrey Keynes, ed., *The Works of Sir Thomas Browne* (London: Faber & Faber, 1931), 5:361.

23. Browne, *Pseudodoxia*, 1:227, 206; in the latter passage, Browne cites Genesis 1:11–12.

24. Robert Sharrock, *The History of the Propagation and Improvement of Vegetables by the Concurrence of Art and Nature* (Oxford, 1672), 5.

25. Highmore, *History of Generation*, 46–50; Sir Kenelm Digby, *A Discourse Concerning the Vegetation of Plants* (London, 1661), 43–44.

26. *Works of Browne*, ed. Keynes, 5:427. As Claire Preston has recently observed, in Browne's work, "Seeds, and seed-like phenomena, lurk everywhere, and the natural world seems to be organized on the basis of germination"

(*Thomas Browne and the Writing of Early Modern Science,* 194). On plants and generation in Browne, see also Egon Stephen Merton, "The Botany of Sir Thomas Browne," *Isis* 47 (1956): 161–71; and Merton, *Science and Imagination in Sir Thomas Browne* (New York: King's Crown Press, 1949), esp. 34–60.

27. Browne, *Pseudodoxia,* 1:83.

28. Pierre Gassendi, for example, argued that rather than accidental, independent actions, what appeared to be instances of spontaneous generation were really the development of hidden seeds created by God (John Farley, *The Spontaneous Generation Controversy from Descartes to Oparin* [Baltimore: Johns Hopkins University Press, 1977], 12).

29. Browne, *Pseudodoxia,* 1:487.

30. Browne, *Pseudodoxia,* 1:207; Browne also suggests that seeds can spontaneously generate "lower" forms of plants, such as "barley into oates" and "wheat into darnell [a much-hated grass]" (1:227).

31. Browne, *Pseudodoxia,* 1:144.

32. *Works of Browne,* ed. Keynes, 5:334.

33. Digby, *Discourse Concerning the Vegetation of Plants,* 79.

34. Peter Sterry, *A Discourse of the Freedom of the Will* (1675), 49, qtd. C. A. Patrides, "'Above Atlas His Shoulders': An Introduction to Sir Thomas Browne," in Browne, *The Major Works,* ed. Patrides, 31.

35. Browne, *Pseudodoxia,* 1:441–42.

36. Merton, *Science and Imagination,* 44. In *Christian Morals,* Browne asserts that God "hath in his Intellect the Ideal Existences of things, and Entities before their Extances" (*Works,* ed. Patrides, 468).

37. Gasking, *Investigations into Generation,* 42.

38. Havenstein, *Democratizing Sir Thomas Browne,* 64. As Havenstein notes, Browne uses the word "woman" or "women" only ten times in *Religio Medici,* six times "as an abstract concept in a discussion of Scripture," and twice in the vicinity of the word "never."

39. Browne, *Pseudodoxia,* 1:207.

40. I am thus qualifying Preston, who interprets Browne's "civility" as modeled on the spontaneous generation of bees (*Thomas Browne and the Writing of Early Modern Science,* 80–81). Ironically, Stanley E. Fish chides Browne for *not* wielding the godlike powers of generation toward which he aspires: "his words are not seeds, spending their lives in salutary and self-consuming effects, but objects, frozen into rhetorical patterns which reflect on the virtuosity of their author" (*Self-Consuming Artifacts: The Experience of Seventeenth-Century Literature* [Berkeley: University of California Press, 1972], 372).

41. In early modern England, Keith Wrightson observes, "The family was the basic unit of residence, of the pooling and distribution of resources for consumption.... Through the family, society reproduced itself; children were

born and reared and property was transmitted from generation to generation. Within the family, individuals found security and identity and the satisfaction of both physical and emotional needs not catered for by other social institutions. The family was fundamental" (*English Society, 1580–1680* [New Brunswick, N.J.: Rutgers University Press, 1982], 66).

42. Howell, *Epistolae*, 91; Digby, *Observations upon "Religio Medici,"* 111–12.

43. John Evelyn, *Sylva, or a Discourse of Forest Trees* (London: Doubleday, 1908), 2:263; Evelyn, *Publick Employment and an Active Life* (London, 1667), 31.

CHAPTER SEVEN

Human Nature
Observing Dutch Brazil

Julie Berger Hochstrasser

This essay is not so much about nature itself, as it is about the *observation* of nature, or more precisely, about certain *depictions* of observation of nature—and in the end, it will also be about what we learn of *human* nature from these remarkable depictions. We begin with a picture—a curious one indeed. The journal of a German soldier employed by the Dutch West India Company in Brazil between 1642 and 1645 contains an illustration that stops one short (fig. 1). With that curious bearded face and the hint of a smile, is it some sort of a wild man? With those grotesque claws and stump of a tail, is it a monster? Seemingly floating downward, with wisps of vegetation that look like seaweed on either side, is it diving underwater? And what is the significance of the musical score below?

The accompanying text is little help: the soldier labels the subject a *Leugart*, but this word does not appear in modern German dictionaries, and sources on seventeenth-century German merely suggest the possible translation of "lion-like."[1] The author offers an alternate term, *Haut*, which he says is Brazilian, but by this he does not mean

155

Figure 1. Caspar Schmalkalden, *Haut*. Pen and ink with wash, 20 x 17 cm, from *Die Wundersamen Reisen des Caspar Schmalkalden nach West- und Ostindien, 1642–1652*, 110. Forschungsbibliothek Gotha, Chart B533.

Portuguese Brazilian (many of his entries provide Portuguese labels, identified as such, as well as German), but rather an Amerindian dialect, for which there is no dictionary in which to seek its meaning.

The carefully penned notes of explanation compound the curiosity:

> It is reported of this HAUT that it intones the six musical notes, (ut, re, mi, fa, sol, la—la, sol, fa, mi, re, ut), correctly ascending and descending—and between each sound, pauses with a "ha" in a pause of a breath or half beat, which is perhaps why the Brazilians call it a Haut.[2]

Figure 2. Caspar Schmalkalden, *Ai* (Sloth). Pen and ink, 20 x 17 cm, from *Wundersamen Reisen*, 109. Forschungsbibliothek Gotha, Chart B533.

Only when we compare this illustration with the preceding entry in the journal (fig. 2) does it become apparent that this is a(nother) kind of sloth, actually meant to be pictured dragging itself across the ground, and the musical score records the remarkable sounds the creature reportedly utters. What looks at first like a relic of some wildly imaginary medieval travel description by the legendary John Mandeville proves to be reasonably careful empiricism. Awkward, unschooled, and even bizarre as the whole entry may first appear—what an extraordinary piece of observation of nature is this!

The journal was recorded by one Caspar Schmalkalden of Gotha, Germany (near Dresden), where his manuscript resides today in the Landesbibliothek. In 1642 Schmalkalden had sailed aboard the "Elephant" from the island of Texel in The Netherlands to Pernambuco in Brazil. He arrived on December 11 after a month-long journey to the Dutch colony, then under the administration of Johan Maurits, count and later prince of Nassau-Siegen (1604–79). Schmalkalden's *Reisebuch* is but one of a truly marvelous (if occasionally perplexing) array of nature studies originating from Dutch Brazil under the reign of Maurits, which together constitute the focus of this essay.[3]

In what follows, one might find a lesson from early modern European history on the difference that is made by *who* does the looking—what a profound impact this has on *what* is seen, and especially, on *how* it is understood. This study of the transfer of information by visual means within the history of natural science offers an extraordinary demonstration of what I shall call the "whisper-down-the-lane" effect—to borrow a metaphor from the parlor game in which something is whispered from player to player, until what was first said is compared with how it comes out at the far end of the line.

The array of nature studies from seventeenth-century Dutch Brazil generated such interest among Europeans that an overwhelming tangle of copies after copies issued forth, with so many variations on themes that authenticity becomes a bewildering parlor game in itself, and the authority of the "eyewitness" image is called seriously into question. Svetlana Alpers and David Freedberg have both variously argued that pictures *are* science, or *can* be—and this material is surely proof of that claim—but if so, it must also be conceded that this science is as idiosyncratic as the people who make the pictures.[4] Tracking Schmalkalden's curious sloth through this maze of imagery is an exercise in sleuthing and common sense, but in the process, it will also provide a survey of the diverse and epic early modern pictorial records of Brazil's natural history.

Our story begins officially with a milestone in the history of natural description: the *Historia Naturalis Brasiliae,* published in 1648 in Leiden and Amsterdam, under the names of Willem Piso (1611–78) and Georg Marcgraf (1610–44).[5] Marcgraf was a German naturalist and cartographer; Piso was also German but was educated as a physician in Leiden. The two were employed by Johan Maurits when he was appointed in 1636 by the Dutch West India Company to serve as governor-general of the new Dutch colony in Brazil.[6]

The *Historia* was the formally commissioned, scholarly fruit of Johan Maurits's patronage in the arena of natural science. While, as we shall see, it was only the tip of the proverbial iceberg, it is eminently worthy of consideration in itself. The book provided 533 woodcut illustrations, with vernacular names and descriptions of some 301 Brazilian plants and 367 Brazilian animals. Piso's section of the volume offered 3 woodcuts of sugar mills, 9 of animals, and 92 of plants, some duplicating Marcgraf's; the naturalist for his part presented 7 woodcuts of native people, 200 of plants, and 222 of animals, birds, insects, and fishes, most of which had never before been described in print.[7] This was a scientific contribution of enormous magnitude.[8] For the multitude of new species reported and described, it remained the authoritative source for one hundred and fifty years; Linnaeus eventually incorporated them into the named species proposed in the tenth and twelfth editions of his *Systema naturae* in 1758 and 1766.[9] The book remained a prime resource well into the nineteenth century.[10]

The very first animal illustrated in the *Historia* section on quadrupeds (Book 6) is a sloth (fig. 3), and the first sloth in Schmalkalden's *Reisebuch* (fig. 2) turns out to match it exactly. Marcgraf's is labeled *Ai* (Brazilian) or *Preguiça* (Portuguese)—names Schmalkalden also lists—and *Ignavus* (Latin) or *Luyaert* (in "our" language, as he says), which clearly relates to Schmalkalden's German *Leugart*.[11] The match with Schmalkalden's first sloth drawing presents a logistical puzzle, however, since the *Historia* was not published until 1648, three years after Schmalkalden had left Brazil and four years before he finally returned to Europe from his subsequent travels throughout Sumatra, Taiwan, and Japan; and it is unlikely that he filled in his sloth drawing after his departure from Brazil, since the text of his manuscript runs continuously down to the top of each drawing, then resumes directly beneath it.[12]

But the sloth published in *Historia Naturalis Brasiliae* turns out to be an exact copy of an earlier illustration in the 1605 *Exoticorum Libri Decem* of Charles l'Ecluse, known in Latin as Carolus Clusius (fig. 4).[13] Such borrowings occur elsewhere in the *Historia*: among the quadrupeds, an armadillo repeats another Clusius illustration, and among the fish, a "piraaca."[14] This raises the possibility that Schmalkalden, too, found his motif in Clusius—except many other images in Schmalkalden's drawings match up with the illustrations in the *Historia Naturalis Brasiliae* but do *not* have precedents in Clusius's work. Still, our smiling sloth (fig. 1) is not among them.

A number of the *Historia* woodcuts are actually relatively crude reproductions of more sophisticated originals produced by several professionally trained artists who were commissioned to accompany Johan Maurits to Dutch Brazil, to work alongside the scientists and scholars on this project. While Maurits makes mention in a 1678 letter of as many as six artists, we know for certain the identity of only two: Frans Post (ca. 1612–80), a landscape painter, and Albert Eckhout (ca. 1610–66),

Figure 3. *Ai* (Sloth). Hand-colored woodcut from Willem Piso and Georg Marcgraf, *Historia Naturalis Brasiliae* (Leiden: Francis Hack; and Amsterdam: Elsevier, 1648), 221. St. Louis, Library of the Missouri Botanical Garden. © 1995–2007 Missouri Botanical Garden http://www.botanicus.org/.

Figure 4. Woodcut from Charles l'Ecluse (Carolus Clusius), *Exoticorum Libri Decem: Quibus Animalium, Plantarum, Aromatum, aliorumque peregrinorum Fructuum historiae describuntur* (Antwerp: Ex Officina Plantiniana Raphelengii, 1605). St. Louis, Library of the Missouri Botanical Garden. © 1995–2007 Missouri Botanical Garden http://www.botanicus.org/.

who specialized in figures, flora, and fauna for the enterprise.[15] Post and Eckhout set sail with Maurits's retinue from the island of Texel in the Dutch Republic, probably with one of the boats that left on October 25, 1636, and arrived in Recife in January 1637, staying until 1644. Together these painters produced what are now the oldest extant pictures of Brazil (or anywhere in South America, for that matter) by trained European artists based on their own eyewitness observation.

Best known among their productions are the oil paintings on canvas, which are fascinating and valuable representations from early modern Brazil.[16] Behind these finished oils, however, lay an array of studies of great beauty and delicacy, which also in some cases served as models for woodcut figures in the *Historia*, and which prompted a proliferation of copies. Post produced a series of landscape drawings in pen and wash that provided source material for his paintings and later graphics for the *Historia* and elsewhere.[17] Eckhout generated dozens of extraordinarily fine studies in oil on paper, of the people, the plants, and the animals of the colony, some of which are related to his renowned life-size oil-painted figures and still lifes, but some also to various *Historia* illustrations.

Johan Maurits brought all this artwork back with him, along with his collected *naturalia,* when he resigned his post as governor and left Brazil in 1644.[18] In 1657 he passed Eckhout's studies along to the Elector of Brandenburg, who had his court physician Christian Mentzel organize the drawings in four bound volumes entitled the *Theatri rerum naturalium Brasiliae,* containing some 417 drawings and oil sketches, principally of animals and plants but also of natives.[19] Since many sketches depicted more than one organism, the total count of depicted items is considerably higher: 569 altogether, including 83 sea creatures in the "Icones Aquatilium"; 115 birds in the "Icones Volatilium"; 73 figures in the "Icones Animalium," including both animal and human subjects; and 298 plants in the "Icones Vegetabilium."[20] Thirty-three additional images of Brazil, including crayon sketches and some studies duplicating *Theatri* subjects, were sent to Mentzel by Andreas Cleyer of Kassel, who had long managed a Dutch trading post in Decima, Japan; collected now as the *Miscellanea Cleyeri,* these are believed to be pieces Mentzel withheld from the *Theatri.*[21]

As if their origins were not colorful enough, the subsequent adventures of these exquisite studies in the *Theatri* are worthy of a Hollywood screenplay. From the Elector's collection they made their way into the Preussische Staatsbibliothek in Berlin, where they formed part of a larger group collectively known as the *Libri Picturati.*[22] These and other valuable documents, including some rare autograph musical manuscripts, were evacuated in 1941 to Lower Silesia for safekeeping from Allied bombing raids.[23] In 1944 they were moved again to Grüssau (today Krzeszów), eventually (to make a long and dramatic tale short) ending up in the Jagiellon Library in Kraców, where Peter Whitehead's determined search was at last rewarded in the late 1970s. The papers were finally made publicly accessible in 1981.[24]

The comments of Frederick's physician introducing the *Theatri* offer insight into the appreciation for these images at the time: "These pictures have given His Serene Highness the Elector supreme pleasure not only because they represent nature's magnificence but also because they constitute a treasure that enriches all erudite curiosity."[25] Later,

Mentzel intimates the familiar adage that one picture is worth a thousand words: "So behold the magnificence of this 'Theatre,' its splendour, its recommendation: that the powers immediately perceive at a single glance the truth and authenticity of the things, without the need for long detailed descriptions." And better still:

> In practice, such greater worth hath painting than eloquence, principally when the reproductions are authentic and created by skilled hands which do not intend to surpass natural beauty, and which His Highness the Prince hath presumed to avoid with utmost diligence. This was his greatest criterion in painting, there being no other endeavor than to conserve the natural disposition of things.[26]

Indeed, most of the drawings in the *Theatri* have the arresting quality of studies *naer het leven* (after life), whether capturing the averted gaze of the shy *Tapiirete* (tapir), flipping over the *Nhamduguaçu* (tarantula) to view its underbody, or locking gazes with the human inhabitants of the colony, from the skeptical sidelong frown of an albino Negro to the pensive close-up visage of a young Tapuya man.[27] Among Eckhout's studies in the "Icones Animalium" is a painted study of a sloth or "Ai" (fig. 5), different from the woodcuts in Marcgraf's *Historia* or Clusius's *Exoticorum*. Rendered in oil on paper, it exhibits the upswept hair that, on the sloth uniquely among all mammals, grows from stomach up toward the spine.[28] Like Schmalkalden's curious sloth, this one is pictured crawling on the ground, and Schmalkalden's pen sketch even suggests a similar mane down the back of its creature. But individual limbs differ in position, and Schmalkalden's view provides information invisible in Eckhout's (the tuft of a tail), while the anatomical accuracy of Eckhout's sloth face bears no resemblance to the disarmingly human features of Schmalkalden's. In short, the differences disqualify this, too, as a potential model for Schmalkalden.

Eckhout's sloth bears a closer resemblance to another important source. In addition to the numerous oil-on-paper studies which found their way into the *Theatri*, a separate but related bunch of watercolors also was passed to the Elector by Maurits. Included in the *Libri Picturati* now in Kraców, they were two volumes, bound as the *Libri Principis* (Books of the Prince), also known today as the *Handbooks*.[29] They seem

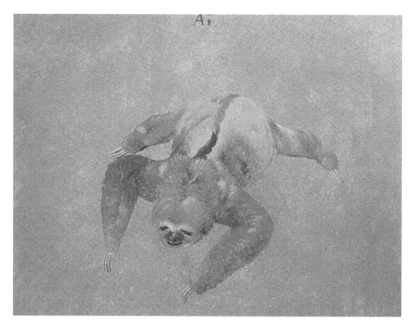

Figure 5. Albert Eckhout, *Ai* (Sloth), ca. 1640. Oil on paper, 50 x 32 cm, from *Theatri Rerum Naturalium Brasiliae, Libri Picturati* A 34, Biblioteka Jagiellonska, Kracόw.

to have been conceived as a private guide or manual for Johan Maurits's own education and leisure, to acquaint him with the animals and plants in his vast dominion in the New World.[30] The *Libri Principis* contained 460 pages of thick-grained white paper, with 93 watercolors and sketches of 37 mammals, 85 birds, 42 fishes, crustaceans, turtles, and other aquatic animals, and 29 insects, arachnids, and reptiles.

The artist of the *Handbook* sketches is not recorded, but they are generally attributed to George Marcgraf, the naturalist of the *Historia Naturalis Brasiliae*.[31] Close inspection reveals that most of the woodcuts in the *Historia* match up precisely in pose and detail with renderings in the *Handbooks*, making it likely that the studies in the prince's handbook provided the models for all but a few of the *Historia*'s prints.[32] Altogether, 22 of the 27 creatures represented in the *Historia*'s section on quadrupeds seem based directly on sketches in the handbooks.[33] Seven of the *Handbook* studies are reversed left-to-right in the *Historia*, as would result from using the exact same drawing for the carving of the woodblock; fifteen others match the orientation of the *Handbook* drawings, indicating a tracing would have been flipped before being transferred to the block—but all are virtually identical in their contours. Similarly, numerous relationships pertain throughout other sections of the *Historia*: lizards, birds, fish, and insects galore reproduce the watercolor studies in the prince's handbook in cruder, but of course reproducible, woodcut versions.[34]

Further support for the attribution of the sketches in the *Handbooks* to Marcgraf is found in the kind of contextual detail they include. Although the renderings in the *Handbooks* are not generally as accomplished or sensitive as those in the *Theatri*, still, they depict specifics of habitat that bring their own convincing air of authenticity to the observations, strongly suggesting the hand of a naturalist in the field.[35] Some captive specimens could have been observed in Maurits's renowned menagerie, but certain sketches bear the mark of animals in the wild: the large anteater is shown with an ants' nest and an adjoining tree trunk crawling with more ants, while a delicate deer is caught in the act of drinking from a trickling spring.[36]

The *Historia* woodcuts generally excerpt the animals from these richer contexts of habitat in the *Handbook* studies: for example, the monkey described as a *cercopiteco* in the *Historia* reproduces precisely the pose captured in the *Handbook* sketch, down to the curl of the long tail around a tree trunk, but the *Historia* eliminates the trees so the tail curls around nothing.[37] Consistent with this tendency, the page in the *Handbooks* devoted to sloths places them in as detailed an environment as depicted anywhere: two creatures are shown, one climbing a tree trunk toward a meal of leaves above, and the other dragging itself along the ground (fig. 6). Neither one bears the slightest resemblance to the standing profile creature represented in the *Historia*, in visual quotation from Clusius—though the creeping sloth does rather resemble Eckhout's, now turned to the observer's left to move horizontally across the green earth beneath the tree.

Many of Schmalkalden's creatures also look as if they have been copied from the *Handbooks*: from the spotted ray with the arc of its tail curving smoothly back around, to the bug-eyed grin of his flying fish with its peculiar fairy wings.[38] But neither of the *Handbooks*' two sloths provides a model for the *Reisebuch*'s: even rotating Marcgraf's horizontal

Figure 6. Attributed to Georg Marcgraf, *Ai* (Sloth). Pen and ink and watercolor, 32.2 x 21.5 cm, from *Libri Principis, Libri Picturati* A 36, Biblioteka Jagiellonska, Kraców.

creeper ninety degrees counterclockwise to orient it with Schmalkalden's curious sloth, the naturalist's sloth face again exhibits nothing of the comically anthropomorphic character of Schmalkalden's.

The written notes in the *Handbooks* do, however, shed light on Schmalkalden's text. Christian Mentzel's preface, quoted earlier, which seems to have been written for both of the collections the Elector had been so fortunate to receive, attests to the eyewitness character of the pictures, which we have just addressed; but it also discusses the notes in the *Libri Principis* (*Handbooks*) as having been penned by Johan Maurits himself:

> Thus, more than any archives, than the abundance of classifications and indices, he [Johan Maurits] esteemed nothing more than to portray the incredible variety of the wondrous world and the nature that in it delighted to dally. In both aforementioned books, which contain the pictures meticulously painted in every detail, the same Prince described in his own hand, in our German language, the true grandeur and dimensions of the creatures portrayed, so that these books must be admired and appreciated as authentic and authorized records of the birds, and land and sea animals. No less authority must be given to the paintings, fruits of the same land, made with the same care and accuracy by those who gazed upon them with their own eyes.[39]

One wonders whether the penned notes are really those of the prince or the naturalist: in proper empirical spirit, Marcgraf had declared, "I will not write about anything which I have not actually seen and observed," so one would expect that if these were his field studies, he would have found it useful to annotate them.[40] Mostly limited to comments about the size and character of the animal, the notes in the *Handbooks* follow a convention familiar from zoological writing and found also in the *Historia*: the impulse to describe unfamiliar fauna and flora through the invocation of more familiar species in the Old World. So, in the *Historia*, a jaguar is described as the size of a wolf; a coati is compared to a cat; and the sloth is *magnitudine mediocris vulpis nostratis*—"the size of our average fox."[41] It should be noted, however, that the observations in the *Historia* do not always reiterate those in the *Handbooks*: for example, the *Handbooks* compare the coati to a fox rather than a cat.

Still, as for the sloth, the entry penned in the *Handbooks* does accord with the *Historia* text in calling the Three-toed Sloth an *Ai* or "Luijaert,"[42] which recalls Schmalkalden's designation of his beast as a *Leugart*—does his handwriting perhaps rather read *Leuyart*, closer still to the other text?[43] Maurits's remarks (or Marcgraf's—both were of German birth) specifically reference German nomenclature: "the

Germans call this animal as a sloth (*Luiaert*) since it walks no more than fifty steps a day. It feeds on old leaves. Its hair is thick, and it reaches about two feet long."[44]

Mentzel's discussion of the notes, quoted above, raises another interesting issue: the impetus behind this whole natural history enterprise. He ascribes lofty aims—and indeed, no one could deny that a sincere thirst for knowledge fueled this powerful artistic and scientific project—but later commentators have been more cynical about its ulterior motives. Teixeira is particularly frank:

> Since the Dutch had sent spies to northeast Brazil a few years before to gather information for their planned invasion, once the area was occupied it would be natural for them to be extremely interested in specialist reports in a wide range of fields as these could provide an accurate portrait of the current situation and future potential of the new territory. These reports were no mere interesting descriptions of exotic animals and plants or "barbarian" peoples. Rather, they were the best, if not only, available source of reasonably trustworthy information. As such, they not only became an essential instrument for tactical and strategic assessment of conquest and colonization, but also an active and sometimes revolutionary component of the colonists' own universe as they were exceptionally greedy for the novelties which surpassed even the wildest imagination.[45]

It is a little hard to imagine the poor sloth playing much of a role in such an ambitious scheme, but the question of motives is worth entertaining. Besides the *Theatri* and the *Libri Principis,* there is a confusing proliferation of copies of such material apart from Schmalkalden's *Reisebuch,* and each introduces a slightly different perspective on the material. The most direct is by another German soldier by the name of Zacharias Wagener. Son of a modest Dresden clergyman, he left Dresden at the age of twenty to seek his fortune, signing on as a mercenary to Brazil in the service of the Dutch West India Company. His talent as a draughtsman was recognized by the governor general, who promoted him to the position of chief notary and then to quartermaster or butler for Maurits himself. In this capacity Wagener created a collection of rather accomplished watercolor drawings, including studies of plants and animals and views of the territory, and what are apparently copies after Eckhout's monumental figures.

Teixeira notes the interest of Wagener's account precisely for its unique point of view: "[the] 'Thierbuch,' written by butler Zacharias Wagener, provides an invaluable contrast to the opinions of those cultured men who pored over the New World's nature for strategic or academic motives." Here Teixeira emphasizes the utilitarianism that "always ran through the Christian West's relations with what was called the 'natural world.'"[46] Certainly, Wagener's duties as Maurits's quartermaster noticeably affect the character of his project: his *Thierbuch* (Animal Book) pays special attention to edible produce: he mentions that guinea pigs, "after being killed, are a tasty delicacy," while of the *Peixe Gate*, he observes: "What we call a catfish, large quantities of these are caught every day and brought in for sale. They have no scales, are completely covered in red and black spots. They taste delicious, pleasant and sweet, which is why almost every afternoon I order some to be taken to His Excellency."[47] But he documents plenty of other nonedible creatures as well, frequently (though not always) copied obviously from the visual precedents in the *Libri Principis*: his deer is also drinking; his possum and agouti are identical though without their settings; and the wild pig is the same drawing in reverse.[48] Sure enough, Wagener's creeping sloth also bears a marked resemblance to Marcgraf's in the *Handbooks*, only extending its arms ever so slightly differently (fig. 7). Perhaps because he must converse with the locals to obtain produce for Maurits, his labels are all in Portuguese; he lists his sloth under the Portuguese *Priguiça*. His comments echo others but are his own:

> This animal is called slothful by our people and by the Portuguese, and is thus given the fitting name for its action. It needs a good day to climb a tree, it hangs there by its three-toed front paw, hanging there for a long time until the next paw, after a good half hour, comes to help; then, finally, it brings its hind legs up, very slowly. It only eats some sweet leaves and grasses and does no harm to anybody.[49]

The functional application for Wagener's observations in cataloguing information about comestibles may also help to explain why his renderings of various fruits and vegetables are particularly strong in

Figure 7. Zacharias Wagener, *Priguiça* (Sloth). Ink and watercolor on paper, ca. 21 x 35.3 cm, from Wagener's *Thierbuch*, 78. Kupferstich-Kabinett, Staatliche Kunstsammlungen Dresden, Ca 226a⁵.

their execution. The studies of flora and fauna in the *Theatri* are presumably from Eckhout's hand, since many were obviously utilized in the oil-on-canvas still lifes he painted to decorate the upper register of the great hall in Maurits's palace of Vrijberg in his capital city Mauritsstad, in present-day Recife. The *Theatri* study of monkeypot fruit, for instance, provides a clear case of such direct transcription to Eckhout's larger still life of the same subject.[50] Yet many of Wagener's studies are fully on a par with those in the *Theatri*, and some even outdo them: his cashews, "Caju" (fig. 8), though related in scale to the version supplied in the *Theatri* (fig. 9), are even more lifelike, if compared with the authentic article, still sold by the basketful along the roadsides on Pernambuco's island of Itamaracá.[51] We know that both Eckhout and Wagener would have had ample opportunities to study such subjects from life, not only since, as Wagener writes, "all of Brazil, so to speak, is covered with this fruit tree," but also since both men would have had access to Maurits's impressive botanical garden, an extraordinary contribution to natural history in itself.[52] Some two thousand adult coconut palms, forty to seventy years old, were carefully uprooted from their natural environment and transported by boat over distances of three to four miles to the island of Antonio Vaz, surrounding Maurits's palace of Vrijberg. Barlaeus recounts the marvelous collection of other species collected as well: orange, lime, pomegranate, and fig trees, papaya, genipap, mangaba, calabash, palm, Surinam cherry, custard apple, jamacuru and banana

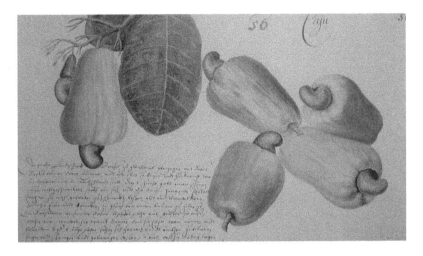

Figure 8. Zacharias Wagener, *Caju* (Cashew). Ink and watercolor on paper, ca. 21 x 35.3 cm, from Wagener's *Thierbuch*, 56. Kupferstich-Kabinett, Staatliche Kunstsammlungen Dresden, Ca 226a[5].

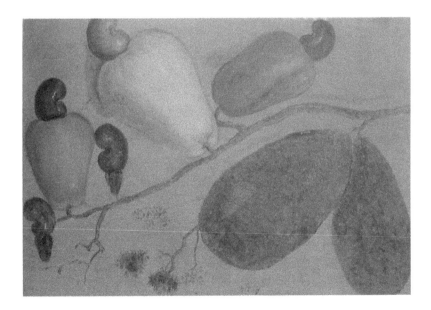

Figure 9. Albert Eckhout, ca. 1640. Oil sketch on paper, 50 x 32 cm, from *Theatri Rerum Naturalium Brasiliae, Libri Picturati* A 35, Biblioteka Jagiellonska, Kraców.

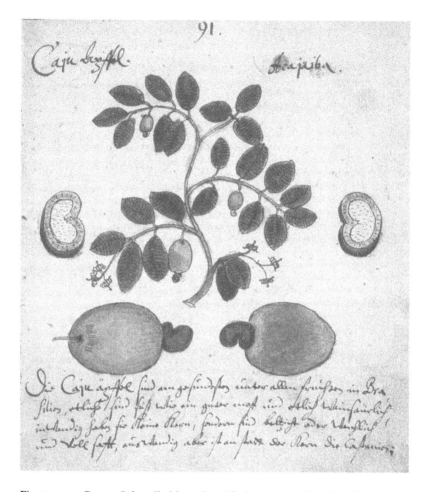

Figure 10. Caspar Schmalkalden, *Caju* (Cashew); *Acajaiba* (Brazilian). Pen and ink and watercolor, 20 x 17 cm, from *Wundersamen Reisen*, 91. Forschungsbibliothek Gotha, Chart B533.

trees, vineyards, a variety of vegetables, and even, besides the native hardwood trees, some from other continents, such as ebony, date palms, and tamarind trees from Africa, and mangoes from Asia.[53]

The close relation between the studies of Eckhout and Wagener is particularly evident when contrasted with the study of the cashew in Schmalkalden's journal (fig. 10). Crude as his sketch may be, it does the careful botanical work of laying out the leaf patterns of the

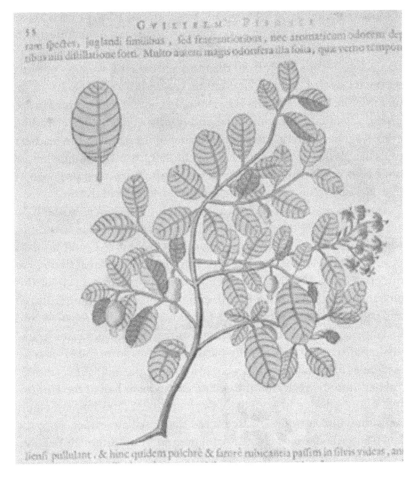

Figure 11. *Acaju* (Cashew). Hand-colored woodcut from Piso and Marcgraf, *Historia Naturalis Brasiliae*, 58. St. Louis, Library of the Missouri Botanical Garden. © 1995–2007 Missouri Botanical Garden http://www.botanicus.org/.

branches, just as does the *Historia Naturalis Brasiliae* (fig. 11). Schmalkalden even goes beyond the *Historia* in providing cross-sections of the nut within the stem of the colorful fruit: like his musical score recording the song of the sloth, this scientific detail is not provided in other contemporary sources.[54]

Schmalkalden's humble journal holds other surprises. Generally speaking, as we have seen, his drawings are put to shame by the more sophisticated hands of Eckhout, the professional artist, or even Marcgraf, the naturalist. On at least one page, however, he trumps them both. A delicately colored drawing of a blue and yellow macaw (fig. 12) stands out so completely from the monochrome ink sketches of birds on the pages that surround it, that the Landesbibliothek Gotha attributes this one drawing to Albert Eckhout. In comparison, the bird in the prince's manual looks emaciated and utterly wooden (fig. 13); while even Wagener's *Arara,* although offering more detail in the feathers of its wings (fig. 14), does not convey the soft feathered volume of the plump body in Schmalkalden's sketch, or the unnerving presence of this bird's piercing gaze, looking the viewer straight in the eye.

Does this reattribution of the macaw in the *Reisebuch* to Eckhout do a disservice to the talents of Schmalkalden, the self-made naturalist, who might just have had his moment of glory here? Could he have profited from close familiarity with a live bird from which to observe directly? Curiously, the *Theatri* collection contains no study of this characteristically Brazilian bird in either its blue and yellow or its red and yellow variants, both of which are illustrated in the *Handbooks* and in Wagener's *Thierbuch*—it includes only a black "Maracana-arara" dissimilar to all of these in the particulars of its form.[55] This may be evidence that some sketches were lost during the collection's moves. Painted into the coffers of the ceiling of the main hall of the Hoflössnitz hunting lodge in Radebeul, Germany, near Dresden, are eighty individual birds obviously based upon the Brazilian studies, since the labels inscribed are in either Portuguese or Tupi. Although Eckhout served as court painter in Dresden from 1652 to 1663, the quality of these renderings does not seem to support their original attribution to Eckhout, and suggests that another artist worked from his drawings, perhaps even under his supervision.[56] One of the Hoflössnitz birds is a red and yellow macaw (fig. 15).[57] The fact that Schmalkalden returned to Germany in 1652, the same year Eckhout took up his post at Dresden (near both Gotha and Radebeul), prompts speculation as to how exactly these motifs might be related.[58]

In any case, it must be conceded that Schmalkalden's *human* figures, while spirited, do not display any of the same confidence of hand as the macaw. In the case of humans, there can be no doubt that Schmalkalden got his inspiration from the models provided originally by Albert Eckhout. These are the most heavily discussed of the pictorial record

Figure 12. Caspar Schmalkalden, perhaps after Albert Eckhout, *Ein West-Indianischer Rabe, Brasilianisch Trarara* (A West-Indian Crow, Brazilian Macaw). Pen and ink and watercolor, 20 x 17 cm, from *Wundersamen Reisen*, 118. Forschungsbibliothek Gotha, Chart B533.

Figure 13. Attributed to Georg Marcgraf, *Indiansche Raab* (West-Indian Crow). Watercolor, 32.2 x 21.5 cm, from *Libri Principis, Libri Picturati* A 36, Biblioteka Jagiellonska, Kraców.

from Dutch Brazil, in recent years at least, and rightly so: Eckhout's magisterial life-size oils on canvas, commissioned by Maurits almost certainly to adorn the main hall of his Vrijberg Palace, are landmarks: the first life-sized paintings by Europeans of non-Europeans.[59] Four couples represent the peoples of the colony: two indigenous Tapuya Indians, a Tupinamba couple, a Negro couple, and a final pair consisting of a mu-

Human Nature 177

Figure 14. Zacharias Wagener, *Arara* (Macaw). Ink and watercolor, ca. 21 x 35.3 cm, from Wagener's *Thierbuch*, 34. Kupferstich-Kabinett, Staatliche Kunstsammlungen Dresden, Ca 226a⁵.

latto man and a mameluca woman.[60] The *Historia* reproduced only the indigenous Tapuya and Tupinamba ("Brazilian") couples, but much was lost in the crude translation to the bookplate woodcuts (see fig. 16, the Tapuya couple).[61] Eckhout's paintings are rich with detail, from the Tapuya man's wooden ear-cylinders, the wooden sticks inserted in his cheeks, and the blue stone placed in his lip on the occasion of his marriage, to his penis sheath of vegetal material; but none of these details show up in the corresponding book illustration.[62] And, like the animals extracted from Marcgraf's nature studies for the *Historia*, Eckhout's humans as transposed to the woodcuts are removed from the articulated landscape settings that help accentuate their character in the paintings through associations with distinctive Brazilian flora and fauna, such as the bird spider and boa constrictor that enhance the savage aspect of Eckhout's armed Tapuya warrior in the corresponding painting.[63]

The *Historia*'s Tapuya woman in the woodcut retains the bunches of leaves with which she only partially conceals her nudity in Eckhout's

Figure 15. After Albert Eckhout, *Arara* (Macaw), ca. 1653–1660. Oil on canvas, 90 x 70 cm. One of eighty birds in ceiling of Hoflössnitz Lodge. Radebeul, Germany, Stiftung Weingutmuseum Hoflössnitz.

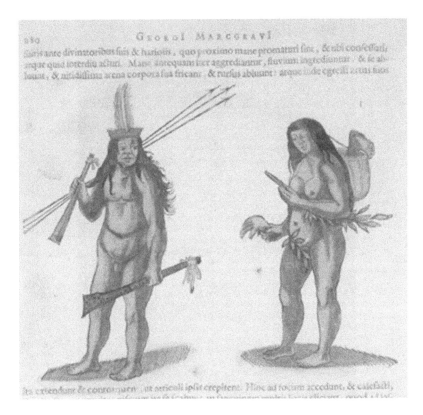

Figure 16. *Tapuya.* Hand-colored woodcut from Piso and Marcgraf, *Historia Naturalis Brasiliae*, 280. St. Louis, Library of the Missouri Botanical Garden. © 1995–2007 Missouri Botanical Garden http://www.botanicus.org/.

painting (fig. 17), and—still more crucially—also the attributes of her alleged cannibalism: in both versions, a severed human foot protrudes from a basket suspended from her forehead, while her own hand grips another, severed human hand. But the *Historia* again has isolated her from Eckhout's jungle setting and the wild landscape in the distance where—just glimpsed between her legs—a hunt is underway. The *Historia* omits the calabash gourd used for eating and drinking that joins the foot in her basket; nor does it capture her unnervingly vacant stare as Eckhout portrays it.[64]

Zacharias Wagener's copies of these figures in his *Thierbuch* meticulously preserve the details of Eckhout's costumes, though the landscape

Figure 17. Albert Eckhout, *Tapuya woman holding a severed hand and carrying a basket containing a severed foot*, 1641. Oil on canvas, 272 x 165 cm. Copenhagen, Nationalmuseet.

settings are much abridged. For instance, the Tapuya woman and the dog, rocks, and water stand against an otherwise blank page (fig. 18).[65] While Wagener's studies of plants and animals show evidence of his own hand at work, sometimes independently of the studies from the *Theatri* or the *Handbooks,* his human subjects show irrefutable evidence of being modeled on Eckhout's figures—or perhaps some now-lost studies for

Figure 18. Zacharias Wagener after Albert Eckhout, *Molher Tapuÿa* (Tapuya Woman), in or before 1641. Watercolor, ca. 21 x 35.3 cm, from Wagener's *Thierbuch*, 90. Kupferstich-Kabinett, Staatliche Kunstsammlungen Dresden, Ca 226a[5].

them. Isolated as they are from Eckhout's backgrounds, Wagener's copies nevertheless depict not only Eckhout's details but also some details that do *not* appear in Eckhout's oils: beside Wagener's African man is a large shield that does not appear in Eckhout's painting, and Wagener's African woman displays a brand on her chest of a crowned "M" that is not visible on Eckhout's woman.[66] Like Post's sunny Brazilian landscapes, which never depict the murderously harsh labor of what one period account called "the hell of the sugarmill," or like his delicately drawn studies of the mills translated into other woodcuts for the *Historia Naturalis Brasiliae* as the very picture of orderly routine, never showing the ax reportedly kept at the ready to chop off the hand of any poor slave who might have gotten caught up in the grinders before his whole body would be drawn in and crushed—do Eckhout's omissions too suggest that he purged his official government version of explicit references to human bondage?[67]

We have some further clue to Wagener's thoughts in the detailed notes that accompany his drawings. While his reference to the practice of branding slaves in his annotation of the "Negro Woman" is chillingly dispassionate, another drawing later in his book has no precedent whatsoever in the painted works of Eckhout or Post, in that it lets slip some hint of human compassion and even perhaps some critical judgment regarding the Dutch slave trade. It depicts slaves being inspected for sale on the Rua dos Judeos in downtown Mauritsstad.[68] His inscription reads: "On the appointed day, these poor people, half dead from hunger and thirst, are taken one by one, as if pigs or sheep leaving the pen, to be counted..."[69] He may be employed by the Dutch West India Company, but Wagener is not Dutch; is this the distance that permits that critical edge?[70]

Like the *Historia*, Schmalkalden reproduces only the two indigenous couples among the four pairs painted by Eckhout. Like Wagener, he labels the "*Tapoijar*" as such but calls the Tupinamba couple simply "Brazilian." His drawing skills fail him visibly in rendering the human figure, differentiating these images notably from whichever models he might have followed. Still, notwithstanding slight variations in pose, the similarities in attire and attributes are too systematic to dismiss. As in Eckhout, the *Historia*, and Wagener, Schmalkalden's Tupinamba woman holds a baby in one arm, balancing a square basket on her head with the other. His Tupinamba and the Tapuya men likewise follow fairly closely the elements that appear in each of the three other versions— all are sufficiently similar to indicate that Schmalkalden was looking at some related model, though again, it is hard to say which.

Where Schmalkalden goes his own way, however, is with his Tapuya woman (fig. 19), who diverges from the Eckhout model to which Wagener and even the *Historia* adhered closely. Now she carries a bird in one hand, and in the other—gone is that severed human hand, replaced by a calabash gourd container. Yet Schmalkalden does not shrink from the thought of cannibalism—on the contrary. The human foot still protrudes from his woman's basket, and still more telling is his accompanying text.

> When the women bear children, so they cut the umbilical cord of the child with a sharp mussel and roast it, with the afterbirth, and eat it all, and when also the child is born dead, they cook

Figure 19. Caspar Schmalkalden, *Tapoÿarisch Weib* (Tapuya Woman). Pen and ink and watercolor, 20 x 17 cm, from *Wundersamen Reisen*, 20. Forschungsbibliothek Gotha, Chart B533.

it the same and eat it, saying, they could not be protected better anywhere. If someone of them dies, the priests come and hew the head and arms and legs off the body, and shred it organ by organ. The wives of the dead or close friends cry and mourn the body, and roast the cut-up pieces. Then the close friends sit together and share the *parentalia*,[71] and therewith nothing is left over, so the old women pick clean the bones with their teeth.

> These bones they save for their great celebration, where they burn them to ashes, mix them in a drink and swallow them, and this happens not out of *desire for revenge*, but rather therewith to demonstrate the great *love* they bear to the dead.[72]

Schmalkalden reports all this with the equanimity of the best modern-day ethnographer, passing no judgment whatsoever. And in fact he has got it exactly right: contrary to the sensationalized accounts of an earlier generation, these people practiced *ritual* cannibalism, *not* directed at hapless travelers.[73] To appreciate just how remarkable Schmalkalden's measured scientific tone was, contrast Wagener's account of the same custom. He, too, gets it essentially correct, but he cannot refrain from expressing how deeply this practice offends his sensibilities: "What is truly horrific, however, and for many ears abominable, is that when a child is stillborn, the mother immediately cuts it up and eats as much as she can...."[74]

Once again, as with that most remarkable drawing with which this essay began, where Schmalkalden diverges from his sources he has the most to offer, through his not inconsiderable innate inclinations as amateur ethnographer and naturalist. The strangeness and uniqueness of the curious sloth drawing suggests Schmalkalden's own hand at work; inexplicably, however, it was left out of Joost's 1983 publication, which transcribes all the German text and reproduces all the *other* images of Schmalkalden's original *Wundersamen Reisen*.[75] The originality of observation by this common German soldier of insignificant rank and his divergences from other sources both pictorial and textual, created by hands more schooled and scholars more learned by far, also serve to make strange the officially sanctioned view of the colony—reminding us, usefully, that those views are themselves neither preordained nor inherently natural. The final testimony to the enduring fascination of these images is the incredibly active life they enjoyed following their return to Europe—stirred up first by Maurits himself.

Soon after Johan Maurits's return to Europe in 1644, he began disseminating the rich treasures of his Brazilian episode in a series of bids to promote his political career.[76] In 1652 Maurits presented the Elector of Brandenburg with the studies of the *Theatri* and the *Handbooks* as described above, along with other treasures, such as carved furniture and the skins of exotic animals. The gift yielded him Freudenberg Castle, near Cleves, and access to noble titles, including that of "Prince." Two years later he made a "grant" to his cousin Frederick III, King of Denmark, which included twenty-six oil paintings, many by Eckhout, probably in exchange for the coveted Order of the White Elephant.[77] He also supplied the University of Leiden with zoological specimens for the famous cabinet of Natural History of Albert Seba, and other private collectors including Frederik Ruijsch and Olaus Worm. And in 1679, shortly before his death, he made another gift, to Louis XIV of France, of paintings by Eckhout and Post, scientific specimens, and "diverse objects and various figures on cardboard," which were later used as motifs for the weaving of the famous Gobelin tapestries series, *Teintures des Indes*—though this final offering failed to yield the returns for which he had hoped.[78]

The tapestries alone represent a significant means for dissemination of the Brazilian imagery. Though the original set was lost, Eckhout had designed them for the Elector of Brandenburg, drawing on the *Theatri* and *Libri Principis* as veritable pattern-books. His designs—the "figures on cardboard" mentioned above, known as cartoons—inspired the *Teintures des Indes* that were woven continuously by the Gobelins until 1730. They remained so popular that even after the original cartoons disintegrated, François Desportes redesigned them from the *Anciennes* version to produce the *Nouvelles Indes* from 1740 through 1786. Dozens of motifs, animal, vegetable, and human, were copied straight from both the *Theatri* and the *Libri Principis* to the tapestries, down to the replication of specific markings and poses.[79]

Thus the original Brazilian work continued to seed yet other proliferations of imagery; but with the successive degrees of separation of its various reworkings, the accuracy and precision of the original records gradually dissipated. The artist Samuel Niedenthal and the naturalist Jacob-Wilhelm Griebe rendered their own fragmentary versions; even John Locke had his servant copy a set of the human figures for him, though he somehow succeeded in mislabeling the Tupinamba man as Amboinese—from the Moluccan Islands, thousands of miles away in East Indonesia.[80] The studies also provided ample material for the numerous panels of Jan van Kessel's painting *Allegory of the Continents*, but there the geographical origins of the flora and fauna become as notoriously jumbled as the human and cultural artifacts depicted.[81] This extraordinary afterlife of the pictorial production from Dutch

Brazil attests clearly to an interest in exotica that was widespread throughout Europe at the time.[82] But the mania eventually drove things out of hand in that inevitable process of "whisper-down-the-lane"—an eighteenth-century Delft tile, for instance, depicts Eckhout's Africans rubbing shoulders with the Chinese.[83]

Yet the genuine scientific contribution of the original material remains.[84] The string of dedications introducing the *Dutch Brazil* series of publications on the early colony witnesses the extent to which Brazilians appreciate the contribution this visual culture has made to their natural history; testimonials in the first volume (1995) are from the Ministry of the Environment,[85] the President of the Instituto Cultural Mauricio de Nassau, the Brazilian Ambassador to the Netherlands,[86] and the list goes on.

A former ambassador of Brazil to the Netherlands notes Maurits's contributions in planning the new city of Recife, including starting "the first zoo and botanical garden, the first astronomy observatory and the first museum of the Americas," as well as in promoting Brazil's image in Europe by disseminating his collections so widely.[87] The Superintendent of the Rio de Janeiro Botanical Garden cites the numerous researchers who credit Marcgraf and Piso's work as "the most important contribution to Natural History since Aristoteles and Pliny."[88] Dante Martins Teixeira, of the Departamento de Vertebrados at the Museu Nacional Universidade Federal do Rio de Janeiro, states, "The value of the collection produced during the so-called 'Mauritian Period,' in what concerns Brazilian Natural History, surpasses every expectation. Furthermore, it constitutes the only comprehensive reference about north-eastern fauna and flora undertaken during a period when the ecosystems were relatively untouched."[89] Summing up both the tragedy of Brazil's environmental losses to date and the value of this early modern legacy, he notes that "the zoological and botanical results of the Dutch occupation of Brazil must be considered as insuperable testimonies of a lost world which has only survived in dispersed forest remnants and in flickering images of fading memories."[90]

The Amazonian rainforest is currently being devoured along with its biodiversity; it is also sobering to be reminded that the Tapuya were all gone by the nineteenth century. Clearly, it is human nature to want to know, and the pictures from Dutch Brazil captured valuable information—about peoples and species, many now extinct, about cul-

Figure 20. Detail from Georg Marcgraf's map *Brasilia qua parte paret Belgis* (Amsterdam: Johan Blaeu, 1657). Engraving, London, British Library. Copyright © The British Library Board, All Rights Reserved, Maps K.A.R. (38.)

tural viewpoints, understandings and misunderstandings, representations as well as misrepresentations. Even Caspar Schmalkalden's funny little drawing of the sloth is a remarkable window onto nature and human nature alike. Shall we conclude that this image is Schmalkalden's original observation, after all?

Minute scrutiny of the vast pictorial array from Brazil turns up an answer of sorts. There *does* appear to be a pictorial precedent for Schmalkalden's image, in a tiny detail in the great map of Brasilia printed on multiple sheets by Johan Blaeu.[91] The map is credited to Georg Marcgraf and dated 1643, but it draws on studies by Post and Eckhout for many of its detailed vignettes.[92] Here in this grand display is one more diminutive sloth (fig. 20), and it is a dead ringer for Schmalkalden's odd design. Though scarcely discernible at this tiny scale, here is the same head-downward pose seen from above as in Schmalkalden's version, showing the mane or stripe down the back, and most suggestively, a face with something very like the distinctly humanoid features of Schmalkalden's sketch. Equally telling are the tufts of foliage at either side of the creeping creature's outstretched forearms, which gave the Schmalkalden sketch the misleading appearance of an underwater scene.

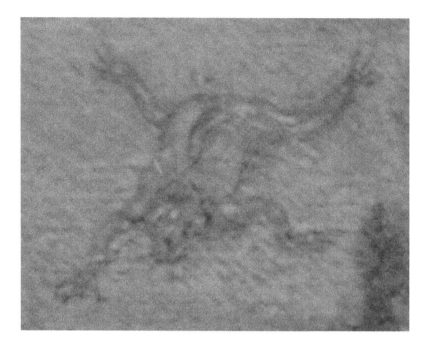

Figure 21. Jan van Kessel, detail from "Potosi" panel of "America," *Allegory of the Continents* (central panel dated 1666; "Potosi" panel oil on copper, 14.5 x 21 cm). Munich, Bayerische Staatsgemäldesammlungen, Alte Pinakothek.

Marcgraf's map is dated 1643; could this possibly mean the naturalist drew this time on Schmalkalden, or did Schmalkalden draw on yet another model, now long lost? By this juncture the only thing entirely clear is the difficulty of ascertaining at such a remove who first uttered what to whom, though of course, given Schmalkalden's other borrowings, the latter is by far the more likely scenario—notwithstanding Schmalkalden's transcription of the sloth's musical utterances, which remains unique in the archives. We do have one further breath of evidence. Along with some fifty creatures sketchily integrated from Marcgraf's *Historia* into Jan van Kessel's *Allegory of the Continents*, we find in the panel "View of Potosi" an unmistakable repetition of this tiny sloth as well—along with a "capybara or water pig" and a "Brazilian mouse" also copied from Marcgraf's map (fig. 21).[93] After tracking such a quirky menagerie of subtly shifting images down such a winding lane, it seems appropriate to end with this faintest of whispers.

Notes

I thank Barbara Haeger and the Center for Medieval and Renaissance Studies at Ohio State University for inviting my lecture, which formed the basis of this essay, and the University of Iowa for the Faculty Scholar Award that funded my research in Brazil.

1. For a transcription of the German original text with illustrations reproduced, see Caspar Schmalkalden, *Die wundersamen Reisen des Caspar Schmalkalden nach West- und Ostindien, 1642–1652* (Leipzig: Brockhaus, 1983), 66. For an English translation, see *Brasil Holandês: The Voyage of Caspar Schmalkalden from Amsterdam to Pernambuco in Brazil*, 2 vols. (Rio de Janeiro: Editora Index, 1998), 2:66.

2. Translation is mine. The sources above (note 1) transcribe the musical syllables as "ut, *ce*, mi . . .," but musician and Harvard librarian John Ranck confirms that while the syllable "ut" did replace "do" at the time, there was no known use of the syllable "ce" in place of "re"; examination of the original text verifies that "re" is more likely the correct transcription.

3. The Brazilian commitment to publications on Dutch Brazil has resulted in fundamental contributions to scholarship, which reproduce the fine images and translate text into English and sometimes also Portuguese. Beginning under the dual title *Brasil-Holandês / Dutch-Brazil*, the first release made public at last, in five volumes, the *Theatrum*, *Libri Principi*, and the *Handbooks* (Rio de Janeiro: Editora Index, 1995). Subsequent volumes were often issued under the English title only: in 1997, *Dutch Brazil*, edited by Cristina Ferrão and Jose Paulo Monteiro Soares, made accessible in three volumes the *Thierbuch* and autobiography of Zacharias Wagener and the pictures in the Hoflössnitz Weinbergschlosschen (Rio de Janeiro: Editora Index, 1997). The next year three more volumes followed, containing part of the Niedenthal collection and the work of Wilhelm Griebe, coordinated by Dante Martins Teixeira (Rio de Janeiro: Editora Index, 1998). In 2000 another three volumes presented the remainder of the Niedenthal collection along with drawings by Frans Post in the British Museum and Cuthbert Pudsey's journal of a residence in Brazil (Petropolis [Brazil]: Editora Index, 2000). A 2002 release, again edited by Ferrão and Soares, includes a first volume with information from Ceará from Georg Marcgraf (June–August 1639) provided by Ernst Van den Boogaart and Rebecca Parker Brienen, and two more by Teixeira, discussing Brazilian imagery in the Gobelin Tapestries of the Indies and Jan van Kessel's *Allegory of the Continents* (Petropolis [Brazil]: Editora Index, 2002). Since I refer to these frequently in the notes below for their reproductions of the visual material, I will henceforth generally refer simply to *Dutch Brazil*, with date of publication and volume number.

4. See Svetlana Alpers, *The Art of Describing: Dutch Art in the Seventeenth Century* (Chicago: University of Chicago Press, 1983), and David Freedberg, *The Eye of the Lynx: Galileo, His Friends, and the Beginnings of Modern Natural History* (Chicago: University of Chicago Press, 2002).

5. Willem Piso, *Historia naturalis Brasiliae* (Leiden: Francis Hack; Amsterdam: Elsevier, 1648). The Library of the Missouri Botanical Gardens has graciously scanned every last page of their hand-colored original edition of this treasure, which can now be accessed in its entirety on their website http://www.illustratedgarden.org/mobot/rarebooks/ (accessed February 2006). So in addition to the thanks I have already extended them for generously permitting me to photograph from it in 2001, I now add these additional enthusiastic words of gratitude.

6. The bibliography on the early Dutch colony is extensive, but for a focused and seminal history, see Charles Ralph Boxer, *The Dutch in Brazil* (Oxford: Clarendon Press, 1957). For the work of Brazilian scholars, see José Antônio Gonsalves de Mello, *Tempo dos Flamengos*, 4th ed. (Recife: Instituto Ricardo Brennand, 2001), and Leonardo Dantas Silva, *Holandeses em Pernambuco, 1630–1654* (Recife: Instituto Ricardo Brennand, 2005).

7. P.J.P. Whitehead and M. Boeseman, *A Portrait of Dutch 17th Century Brazil: Animals, Plants and People by the Artists of Johan Maurits of Nassau* (Amsterdam, Oxford, New York: North-Holland Publishing, 1989), 28; Whitehead and Boeseman's work provided an invaluable early overview of the related imagery reviewed in this essay, much of which is now finally published in the Editora Index volumes on Dutch Brazil. See also Boxer, *Dutch in Brazil*, 151.

8. American scientist E.W. Gudger has called it "probably the most important work on natural history after the revival of learning, and, until the explorations of the prince of Neuwied were made known, certainly the most important work on Brazil." Gudger, "George Marcgraf, the First Student of American Natural History," *The Popular Science Monthly* (September 1912): 250–74, quoted in Boxer, *Dutch in Brazil*, 154. Earlier Boxer cites Gudger's remark that "if Marcgraf had lived to publish more of his work, he might well have become the greatest naturalist since Aristotle." Boxer, *Dutch in Brazil*, 150 and n. 4. See also A. Taunay, ed., *Jorge Marcgraf: Historia natural do Brasil. Edição do Museo Paulista* (Sao Paulo: Imprensa Oficial do Estado, 1942), and more recently, Van den Boogaart and Brienen, *Dutch Brazil* 2002, vol. 1, with bibliography.

9. Whitehead and Boeseman, *Portrait*, 27.

10. Boxer recounted its significance as follows: "This work contains the first truly scientific study of the fauna and flora of Brazil, a description of the geography and meteorology of Pernambuco, including daily wind and rainfall records, and an ethnographical survey of the local Amerindian races." Citing the illustrations and descriptions, he adds that Piso's contributions on Bra-

zilian medicine "remained an authoritative work on tropical medicine and hygiene until well into the nineteenth century." Boxer, *Dutch in Brazil*, 150–53.

11. Piso, *Historia*, 221; compare Schmalkalden, *Voyage*, in *Brasil Holandês* 1998, 2:65. Though Marcgraf's native language is German, "Luyaert" appears to be a translation into the Dutch of the West India Company, in whose employ he conducted this research. The word does not, however, appear in the *Woordenboek der Nederlandse Taal*.

12. For the later chapters of Schmalkalden's adventures, in the Far East, see Schmalkalden, *Wundersamen Reisen*, and more recently, Caspar Schmalkalden, *Mit Kompass und Kanonen: Abenteuerliche Reisen nach Brasilien und Fernost, 1642–1652*, ed. Wolfgang Joost (Stuttgart: Thienemann, 2002).

13. Charles l'Ecluse (Carolus Clusius), *Exoticorum Libri Decem: Quibus Animalium, Plantarum, Aromatum, aliorumque peregrinorum Fructuum historiae describuntur* (Antwerp: Ex Officina Plantiniana Raphelengii, 1605). Clusius in turn cites Konrad Gesner, *Historiae Animalium*, 5 vols. (Zurich: C. Froschoverum, 1551–58), which was considered to be the first modern zoological work; it was the most popular of sixteenth-century natural histories, visually cataloguing and attempting to classify all known beasts.

On Clusius see Florike Egmond, "Clusius, Cluyt, Saint Omer: The Origins of the Sixteenth-Century Botanical and Zoological Watercolours in Libri Picturati A. 16–30" *Nuncius* 20 (2005): 11–67, and Florike Egmond, Paul G. Hoftijzer, and Robert P.W. Visser, eds., *Carolus Clusius in a New Context: Towards a Cultural History of a Renaissance Naturalist* (Amsterdam: Edita Publishing House of the Royal Dutch Academy, 2007).

14. The *Historia* introduces the armadillo with Clusius's stylized sixteenth-century print (Piso, *Historia*, 231), rather than Marcgraf's more convincing life drawing in the *Libri Principis* (*Dutch Brazil* 1995, 2:3), but the following page reproduces Marcgraf's *Handbook* study for another type of armadillo which includes the creature rolled up in a ball (*Historia*, 232, cf. *Dutch Brazil* 1995, 2:38). See also *Dutch Brazil* 2002, in which Teixeira has meticulously identified numerous borrowings of Marcgraf's motifs in the works of Jan van Kessel; for the comparisons with L'Ecluse, 3:135.

15. For a brief overview of their painted production, see Rebecca Parker Brienen, "Albert Eckhout and Frans Post: Two Dutch Artists in Colonial Brazil," in *Brazil: Body and Soul*, ed. Edward J. Sullivan (New York: Guggenheim Museum, 2001), 62–99. The letter is dated 21 December 1678, to the Marquis of Pomponne, mentioning "dans mon service le temps de ma demeure au Bresil, six peintres." See *Dutch Brazil* 2002, 2:97 n. 15.

16. This even though it has been proven that Eckhout's paintings take ethnographic license; see Rebecca Parker Brienen, *Visions of Savage Paradise: Albert Eckhout, Court Painter in Colonial Dutch Brazil* (Amsterdam: Amsterdam University Press, 2006). For the special exhibition that temporarily returned

Eckhout's paintings from Copenhagen to the Mauritshuis, see Quentin Buvelot, ed., *Albert Eckhout: A Dutch Artist in Brazil*, ex. cat. Mauritshuis, The Hague (Zwolle: Waanders Publishers, 2004). For an impressive private collection of Post's Brazilian landscapes and related material, see Bia Corrêa do Lago, ed., *Frans Post e o Brasil Holandês na coleção do Instituto Ricardo Brennand / Frans Post and Dutch Brazil in the Collection of Instituto Ricardo Brennand*, ex. cat. Instituto Ricardo Brennand (Recife: Ricardo Brennand, 2003).

17. For example, a drawing by Post in the Rijksprentenkabinett Amsterdam provided the model for the cruder woodcut of an ox-driven sugar mill in the *Historia*; his drawings for the engravings in Caspar Barlaeus's *Rerum per octennium in Brasilia* (1647) are now in the British Museum; see Leonardo Dantas Silva, *Dutch Brazil* 2000, vol. 1: *Frans Post, The British Museum Drawings*.

18. Frustrated by the inadequate military resources the West India Company provided, Maurits tendered his resignation to the directors of the West India Company in hopes of securing better support in response to his threat. Instead, they called his bluff, accepting his resignation and obliging him to (reluctantly) depart. On Maurits's termination as governor, see *Dutch Brazil* 1995, 1:46; Buvelot, *Albert Eckhout*, 132; or for a fuller account of his celebrated departure, see Boxer, *Dutch in Brazil*, 156–57.

19. Duparc cites a total of 419 originals in the *Theatri*; see *Dutch Brazil* 1995, vol. 1. Its illustrations have now been reproduced in *Dutch Brazil* 1995, vols. 4–5.

20. Duparc, *Dutch Brazil* 1995, 1:122 n. 32.

21. On Cleyer see comments made by Sousa Leão cited in *Dutch Brazil* 1995, 1:156. Mentzel explains that there were some chalk drawings which he left out of the bound *Theatri* because the Prince had requested that the differing media be sorted accordingly; see *Dutch Brazil* 2005, 1:2.2.3, "The Miscellanea Cleyeri," 156; and 186 ff.

22. The *Theatri*, the *Libri Principis* (discussed below), and the *Miscellanea Cleyeri* are together now in seven volumes which form part of a larger group collectively known as the *Libri Picturati*: the *Theatrum rerum naturalium Brasiliae* constitute *Libr. Pict.* A 32–35; the *Handbooks* are A 36–37, and the *Miscellanea Cleyeri* form A 38.

23. The musical manuscripts included scores by Bach, Mozart, Beethoven, Schumann, Schubert, and Brahms.

24. For Peter Whitehead's dramatic account of his search for these manuscripts, see Whitehead and Boeseman, *Portrait*. For the behind-the-scenes report from Poland, see Jan Piroźyński, "The Berlin Collection in the Jaguelon Library (the "Berlinka")," in *Dutch Brazil* 1995, 1:83–88.

25. *Dutch Brazil* 1995, 3:18.

26. Ibid., 3:22.

27. *Dutch Brazil* 1995, 5:39 (tapir); 5:60 (tarantula); 5:12 (albino); 5:19 (Tapuya).
28. Family: Bradypodidae. Genus: Bradypus (Three-toed Sloth). Species: 1. True three-toed sloths or ai *(Bradypus tridactylus)*. Lives in areas from Central America to Northern Argentina. 2. Brown throated three-toed sloth *(Bradypus variegatus)*. Lives in areas from Guatemala to Honduras. 3. Maned Sloth *(Bradypus torquatus)*. Only found in Eastern Brazil. The most rare of the five species. Bernard Grzimek, *Grzimek's Encyclopedia of Mammals*, vol. 2, ed. Sybil P. Parker (New York: McGraw-Hill, 1990).
29. These form volumes A 36–37 of the *Libri Picturati*. See *Dutch Brazil* 1995, vols. 2–3.
30. Dante Martins Teixeira, in *Dutch Brazil* 1995, vol. 2, "Notes from the Scientific Editor," preceding n. 1.
31. Whitehead and Boeseman opine that it is tempting to attribute the *Handbooks* to Marcgraf; see Whitehead and Boeseman, *Portrait*. Authorities dispute some attributions, however: Teixeira ventures that variations in style and quality suggest the work of more than one hand, and some images may well have been executed by Eckhout *(Dutch Brazil* 1995, vol. 3); Duparc counters: "The watercolours of Brazilian plants and animals in the two *Handbooks (Libr. Pict.* A 36–37) are certainly not the work of Eckhout" *(Dutch Brazil* 1995, vol. 1).
32. The exceptions prove the rule: besides the two quadrupeds copied from Clusius (sloth and armadillo), only three others seem to be based on depictions among the *Theatri* studies instead: a rabbit *(Historia*, 224—cf. *Theatri*, vol. 3, *Icones Animalium*, in *Dutch Brazil* 1995, 5:30, reversed); a tapir *(tapiirete) (Historia*, 229—cf. *Dutch Brazil* 1995, 5:39, reversed); and a yellow-headed creature labeled *Eirara* in the *Theatri*, and *Ilya and Carigueiba* in *Historia*, 234; two varieties appear in the *Theatri* that resemble it closely (not reversed), though not exactly *(Dutch Brazil* 1995, 5:31). One also has no way of knowing what pages might have been lost from either collection in the intervening centuries, though the *Libri Principis* pages do bear numbers.
33. They are as follows (designating the page number in the *Historia Naturalis Brasiliae* as "H" and the page number in volume 1 of the *Libri Principis* as reproduced in *Dutch Brazil* 1995, vol. 2, as "LP"): *Carigueya* or *Tajibi*, an opossum (H222—LP41, not reversed but without his tree); guinea pig (H224—LP9, reversed and with spots nearly but not identical); paca (H224—LP30, reversed); agouti (H223—LP26, reversed); *Aperea*, a field-rat or bush-rat (H223—LP36, not reversed but without his mound and hole); anteaters, both large (with tree trunk) (H225—LP27, not reversed), and small (H226—LP22, not reversed); monkeys, including the *Guariba* (H226—LP23, reversed and without the tree), *Cagui* (H227—LP17, not reversed), marmoset (H227—LP19, not reversed), and the *Circopithecus* from Guinea and the Congo discussed below

(H228—LP35, not reversed); coati (H228—LP15, not reversed); Tajacu pig from Guinea (H229—LP7, reversed); wild pig from Guinea (H230—LP6, not reversed); and capybara (H230—LP4, not reversed). The abovementioned Clusius armadillo is followed by another which *is* based on a *Handbook* sketch that includes a second creature rolled up in a ball (H232—LP38, not reversed). A porcupine perched on a tree branch also definitely matches up (H233—LP5, not reversed), as do the goat from Guinea (H234—LP28, reversed), jaguar (H235—LP21, reversed), and puma (H235—LP16, reversed).

34. A few of the images of the *Libri Principis* also match up with ones in the *Theatri*: the puma or *Çuguaçuarána*, delicately brushed in the *Theatri*, reappears rendered in the pen and ink and watercolor of the *Handbook* studies under the same label, described as "a small kind of tiger," in a precise echo of the seated profile pose of the *Theatri* specimen (not reversed), with only a curl of the tail differentiating the two (*Dutch Brazil* 1995, 5:32, cf. *Dutch Brazil* 1995, 2:16). This is far from the best of the *Theatri* studies, however; that curled tail is rendered with only two strokes, a dark on one side and a light on the other— far more summarily than, say, the finely cross-hatched tail of the opossum or *Taibi*, giving rise to the possibility that Marcgraf or some other artist besides Eckhout produced the *Theatri* puma (*Dutch Brazil* 1995, 5:27). Such inconsistencies abound throughout the collections, as witnessed below in analyses of studies of the cashew and the macaw.

35. A document recently discovered to be connected with Marcgraf by Brienen and Van den Boogaart sheds new light on his fieldwork; see *Dutch Brazil* 2002, vol. 1. Found among Johan Maurits's archives in the Hague, it tersely records the progress of a slaving expedition in Ceará, which Marcgraf may have accompanied.

36. On Maurits's menagerie, see Boxer, *Dutch in Brazil*, 116; for Marcgraf's field studies as mentioned, see *Dutch Brazil* 1995, 2:27 (anteater); 2:33 (deer).

37. *Dutch Brazil* 1995, 2:35, cf. *Historia*, 228.

38. *Dutch Brazil* 1995 (*Libri Principis*), 2:125; cf. Schmalkalden, *Wundersamen Reisen*, 107 (ray), and LP147—S109 (flying fish).

39. *Dutch Brazil* 1995, 3:22.

40. Cited in Boxer, *Dutch in Brazil*, 154.

41. Piso, *Historia*, 235 (jaguar); 228 (coati); 221 (sloth). Assuming the comments to be Maurits's, Teixeira too has noted this tendency: "Of course Nassau's statements could not avoid the comparison of the New World animals with the European ones . . .," so he calls the coati a fox, or macaws crows, etc. Teixeira observes, moreover, that even today, the German language retains vestiges of this habit, calling the guinea pig (the English nomenclature itself, of course, also being an example of the phenomenon) a "Meerschweinchen," or "piglet from (over)seas," like the "Meerkatze," or "cat from (over)seas." *Dutch Brazil* 1995, 2:179 n. 8.

42. Teixeira transcribes the Marcgraf text as "Leupart" or "Luijaert," noting that these names "could be translated as 'lazy' or 'of a lazy nature,' probably being old forms of the modern 'luiaard' used by the Dutch to name this typical New World animal." *Dutch Brazil* 1995, 2:180 n. 25.

43. Joost transcribes Schmalkalden's as "Leugart," but in light of Marcgraf's original one wonders whether it could be read as "Leuyart" or "Leuijart," still closer to the modern Dutch. *Dutch Brazil* 1995, vol. 2; compare Schmalkalden, *Wundersamen Reisen*.

44. *Dutch Brazil* 1995, 2:34.

45. Teixeira, in *Dutch Brazil* 1997, 2:7.

46. Ibid. 2:8, listing bibliography on this topic: F. Turner, *Beyond Geography: The Western Spirit against Wilderness* (New Brunswick, N.J.: Rutgers University Press, 1983); K. Thomas, *Man and the Natural World: Changing Attitudes in England, 1500–1800* (Harmondsworth: Penguin Books, 1983); A.W. Crosby, *Ecological Imperialism: The Biological Expansion of Europe, 900–1900* (Cambridge: Cambridge University Press, 1986); J. Perlim, *História das florestas: A importancia da Madeira no desenvolvimento da civilização* (Rio de Janeiro: Editora Imago, 1992).

47. Wagener, *Thierbuch*, in *Dutch Brazil* 1997, 2:139 (guinea pig) and 2:41 (catfish).

48. LP pages are *Libri Principis* in *Dutch Brazil* 1995, vol. 2; W pages are Wagener, *Thierbuch* in *Dutch Brazil* 1997, vol. 2: compare deer LP33—W137; possum LP41—W121; agouti LP26—W123; and pig LP7—W135.

49. Wagener, *Thierbuch*, *Dutch Brazil* 1997, 2:138.

50. Compare *Dutch Brazil* 1995, 5:86, with Eckhout, *Still Life with Monkeypot Fruit*, ca. 1640 (canvas, 86 x 92 cm.), Copenhagen, Nationalmuseet, reproduced in Buvelot, *Albert Eckhout*, 97.

51. Compare Wagener, *Thierbuch*, *Dutch Brazil* 1997, 2:103, with *Theatri*, *Dutch Brazil* 1995, 5:82.

52. Sergio de Almeida Bruni, superintendent of the Rio de Janeiro Botanical Garden, hails Nassau's garden at Friburg Palace as "the first botanical collection in Brazil originating from scientific practices" and "a pioneer project in the concept of gardens at that time and which is the first colonial garden to be inserted into a landscape design." *Dutch Brazil* 1995, 1:19–22.

53. See Caspar Barlaeus, *Rerum per octennium in Brasilia et alibi nuper gestarum sub. Praefectura illustrissimi Comitis I. Mavritii, Nassoviae & Comitis, nunc Vesaliae Gubernatoris & Equitatus Faederatorum Belgii Ordd. Sub Avriaco Ductoris Historia* (Amsterdam: Johan Blaeu, 1647).

54. Piso, *Historia*, 158.

55. Compare Marcgraf, *Libri Principis*, in *Dutch Brazil* 1995, 2:97–98; Wagener, *Thierbuch*, in *Dutch Brazil* 1997, 2:69–70; and Eckhout, *Theatri*, in *Dutch Brazil* 1995, 4:166. Unaccountably, there is no *arara* in the *Historia* either; the closest in appearance is a "Perroquet."

56. For a summary of the attribution history of these ceiling paintings, see Brienen, *Visions of Savage Paradise*, 42–43. As possible alternatives to explain the cruder appearance of some of these pictures, she adds both their variable condition, incorporating restorations and overpaintings over the years, and the fact that, as ceiling decoration, these images would not have had to be precisely rendered in order to be read from a distance below.

57. On the pictures at Hofflössnitz Lodge, see also *Dutch Brazil* 1997, vol. 3; and Buvelot, *Albert Eckhout*, 23. As an example of the endless history of copies, in 2004 I ran across this same depiction of a red and yellow macaw reproduced on the wall of the dining room in the Hotel Krasnopolsky in Paramaribo, Suriname.

58. Did Eckhout contribute a drawing to Schmalkalden's journal, as this superb study has clearly suggested to some? But if so, when? In Brazil, or back in Germany? Or did a sheet now lost from the *Theatri* inspire Schmalkalden's surprising flash of brilliance? And where is the *Arara* study behind the depiction on the ceiling of the hunting lodge?

59. Brienen has worked to establish their original placement at Vrijburg, proposing a plan for their arrangement in the hall; see originally Rebecca Parker Brienen, "Albert Eckhout's Paintings and the Vrijburg Palace in Dutch Brazil," in *Albert Eckhout volta ao Brasil / Albert Eckhout Returns to Brazil 1644–2002* (Copenhagen: Nationalmuseet, 2002), 81–91; and most recently in her 2006 book *Visions of Savage Paradise*, which makes an essential contribution to the scholarship as the first book-length monograph on Eckhout's work. Ernst Van den Boogaart first argued that the couples represent a three-tiered hierarchy of civility; see Van den Boogaart, "Infernal Allies: The Dutch West India Company and the Tarairiu 1631–1654," in Van den Boogaart, ed., *Johan Maurits van Nassau-Siegen 1604–1679: A Humanist Prince in Europe and Brazil: Essays on the Occasion of the Tercentenary of His Death* (The Hague: Johan Maurits van Nassau Stichting, 1979), 519–38, and subsequent work, summarized in Rebecca Parker Brienen, "Albert Eckhout's Paintings of the '*wilde natien*' of Brazil and Africa," in *Picturing the Exotic 1550–1950: Peasants and Outlandish Peoples in Netherlandish Art,* Nederlands Kunsthistorisch Jaarboek / Netherlands Yearbook for History of Art 53 (2002), 107–31.

60. A mulatto was the child of a black mother and a Portuguese father, while a mameluca was the offspring of a Brazilian woman and a Portuguese or Dutch man. Brienen presents persuasive evidence that the Negro woman is a slave from Angola, while the Negro man is a trader from the Gold Coast of Guinea, actually represented on African soil. See Brienen, "Albert Eckhout's Paintings of the '*wilde natien*,'" esp. 122.

61. Teixeira acclaims the significance of Eckhout's figures and in particular the Tapuya couple: "Eckhout portrayed Negroes and Indians as human beings of great majesty and manliness, whereas his contemporaries placed them in roles comparable to that of the beasts of burden in their paintings. It is there-

fore not surprising that these are the sole paintings of 'Tapuias' of Northeastern Brazil, and the first illustrations of South American Indigenous people carefully produced by a European artist." *Dutch Brazil* 1995, 1:183 n. 94.

62. For a summary of the debate over the status of Eckhout's representations as "ethnographic portraits," see Brienen, "Albert Eckhout's Paintings of the '*wilde natien*.'"

63. The precise identification of all these details is usefully catalogued in Buvelot, *Albert Eckhout*, 68, where other such expressive congruencies are noted: for example, the domestic character of the mameluca is accentuated by a docile little guinea pig.

64. Another massive life-sized oil on canvas by Eckhout depicts a Tapuya dance, in which the statuesque poses of the iconic couple are replaced with animated gesticulations. While the warlike, fearsome, and primitive qualities of the Tapuya are the subject of repeated comments by European observers, still, Johan Maurits prided himself on establishing a relationship with the tribe: descriptions report another large-scale painting by Eckhout, now lost, depicting the governor general himself among the Tapuya. See Van den Boogaart, "Infernal Allies."

65. Wagener, *Thierbuch*, in *Dutch Brazil* 1997, 2:171.

66. For photographic details that make this clear, see Brienen, *Visions of Savage Paradise*, pls. 2–3, pp. 210–11.

67. See Julie Berger Hochstrasser, "Visual Legacies of Slavery in Suriname? The Stakes of Not Seeing," in proceedings of the conference "Globalization, Diaspora and Identity Formation," Paramaribo, Suriname, 2004, and Brienen, "Albert Eckhout and Frans Post."

68. The street known in the seventeenth century as the "Rua dos Judeos" (Street of the Jews) is today the Rua do Bom Jesus in the historical quarter of Recife.

69. Wagener, *Thierbuch*, in *Dutch Brazil* 1997, 2:195–97.

70. On the tendency for outsiders to criticize more freely, and insiders to sanitize the *pictorial* record, see Hochstrasser, "Visual Legacies of Slavery."

71. Schmalkalden, *Wundersamen Reisen*, 20. Joost indicates that *parentalia* is the Latin word for the ritual honoring the ancestors, a learned reference that indicates that Schmalkalden must have had some education. The only other passage in which he reverts to Latin is the one in which he discusses the penis sheath of the Tapuya man.

72. Translation is my own. German original in Schmalkalden, *Wundersamen Reisen*, 21a, 23. For another English translation, see that of Prof. Álvaro Alfredo Bragança Júnior in *Brasil Holandês* 1998, 1:60. Cannibalism is mentioned by other chroniclers as well.

73. Mark Meeuwse of the University of Winnipeg confirms that it was the Tupi who would eat their enemies, while the Tapuya practiced only ritual cannibalism (in conversation at the Columbia University workshop "The Dutch

Golden Age and the World," March 2007, where he presented the paper "Allies and Subjects: The Legal and Political Status of the Tupi Indians in Dutch Brazil, 1630–1654"). The precedent for sensationalist accounts of Brazilian cannibalism was set by Hans Staden, a German traveler to Brazil who recounts in lurid detail his experience of being captured by cannibals who intended to eat him but from whom he ultimately escaped. See Hans Staden, *Wahrhaftiger Historia und beschreibung einer Landtschaft der Wilden Nacketen Grimmigen Menschenfresser Leuthen in der Newenwelt America* (Truthful History and Description of a Landscape of Wild, Naked, Cruel Man-Eating People in the New World of America), first published in 1557 but which soon became a best seller. See also Hugh Honour, *The New Golden Land: European Images of America from the Discoveries to the Present Time* (New York: Pantheon Books, 1975), 64.

74. Wagener, *Thierbuch*, in *Dutch Brazil* 1997, 2:169.

75. It is, however, reproduced in Texeira's English translation edition of Schmalkalden's journeys, *Dutch Brazil* 1998, 2:67, with translation of accompanying text, 66.

76. His strategic bequests to various courts of Europe deployed the dynamics analyzed in a classic work of anthropology; see Marcel Mauss, *The Gift: The Form and Reason for Exchange in Archaic Societies*, trans. W.D. Halls (New York: W.W. Norton, 2000).

77. Brienen asserts that there is no documentary evidence for this connection, however; see Brienen, "Albert Eckhout and Frans Post."

78. Teixeira, *Dutch Brazil* 1995, 1:118 n. 18, with bibliography on the gradual dispersal of this material.

79. For exhaustive exposition of these visual quotations, see Teixeira, *Dutch Brazil* 2002, vol. 2.

80. For the Niedenthal Collection, with 24 of its 423 illustrations related to Dutch Brazil, including some originals attributed to Eckhout, see *Dutch Brazil* 1998, vol. 1, and *Dutch Brazil* 2000, vol. 2; see also *Dutch Brazil* 1998, vols. 1 and 2, for the mysteriously anonymous (but clearly also related) "Animaux et Oiseaux," probably produced in Germany between 1652 and 1685. On Griebe, see *Dutch Brazil* 1998, vols. 1 and 3, "Naturalien-Buch," by Jacob Wilhelm Griebe, with 50 of its 612 illustrations related to Dutch Brazil. On Locke's copies after Wagener or Eckhout, see Whitehead and Boeseman, *Portrait*.

81. See *Dutch Brazil* 2002, vol. 3: Dante Martins Teixeira, *The "Allegory of the Continents" by Jan van Kessel "The Elder" (1626–1679): A Seventeenth-Century View of the Fauna in the Four Corners of the Earth.*

82. On Dutch agency in playing to this fascination for the foreign, see Benjamin Schmidt, "Inventing Exoticism: The Project of Dutch Geography and the Marketing of the World, circa 1700," in *Merchants and Marvels: Commerce, Science and Art in Early Modern Europe*, ed. Pamela Smith and Paula Findlen (New York: Routledge, 2002), 347–69.

83. The tile is in the collection of the Rijksmuseum, Amsterdam, ca. 1700, 170 x 79 cm. For a reproduction in color, see Whitehead and Boeseman, *Portrait*, pl. 88, p. 321.

84. A German publication from 1938 (just a few years before the *Libri Picturati* would be evacuated from the Staatsbibliothek) does no better, however: on one page, it puts the Tupinamba man with the mameluca, then the Tapuya man with the Tupinamba woman! See Emil Helfferich, *Prinz Johann Moritz von Nassau-Siegen und die Niederländischen Kolonien in Brasilien* (Berlin: Verlag Max Staercke, 1938). Peter Mason has explored the phenomenon of the exotic as exactly this kind of decontextualization of cultural detail; see Mason, *Infelicities: Representations of the Exotic* (Baltimore: Johns Hopkins University Press, 1998).

85. In "The Picture of Biodiversity," Executive Secretary Aspásia Camargo admonishes: "Today when we are reaching the extreme limits of our indescribable sources and resources, we should render homage to those who made the effort of registering, from the beginning, the vigorous growth and the greatness of the Brazilian megabiodiversity." *Dutch Brazil* 1995, 1:9.

86. Irene Perbal Bastin, ibid., 11–12; Ambassador João Augusto de Medicis hails the pictures for combining "priceless scientific value with . . . extraordinary artistic quality." Ibid., 13–14.

87. Former ambassador Affonso Arinos de Mello-Franco, ibid., 15–18. The claim of the first astronomical observatory in the Americas, however (first tendered by Boxer in *Dutch in Brazil*), must be corrected: this may be the first *European* astronomical observatory in the Americas, but the Mayan observatory Caracol, with windows that align with important astronomical sightlines, was built at Chichén Itzá during the Toltec period of its reoccupation shortly after 1000 C.E., thus predating the efforts at Dutch Brazil by at least half a millennium.

88. Sergio de Almeida Bruni, in *Dutch Brazil* 1995, 1:19–22.

89. Ibid., 164. Teixeira explains that another ornithological expedition worthy of the name would not take place until the end of the nineteenth century, by which time the destruction of the environment had already begun.

90. Ibid.

91. George Marcgraf, *Brasilia qua parte paret Belgis* (Amsterdam: Johan Blaeu, 1657). (There were other printings both earlier and later, but this is the British Museum copy from which the detail is reproduced.)

92. For example, a Post drawing for the sugar mill in one vignette still survives complete with the grid pattern superimposed to transfer it: Frans Post, *Sugar Mill*, ca. 1640 (pen and ink, 14.3 x 28.2 cm. Brussels, Musées Royaux des Beaux-Arts), reproduced in Buvelot, *Albert Eckhout*, fig. 2b, 134.

93. *Dutch Brazil* 2002, 3:136: "ai ofte Luÿart" (fig. 194); also "capÿbara ofte water Verken" (fig. 193) and "Brasiliaense muis" (fig. 195).

Bibliography

Alan of Lille. "De Incarnatione Christi." PL 210, col. 579A.
Albala, Ken. *Eating Right in the Renaissance*. Berkeley: University of California Press, 2001.
Al-Biruni. *The Determination of the Coordinates of Positions for the Correction of Distances between Cities*. Edited and translated by Jamil Ali. Beirut: American University of Beirut, 1967.
Alpers, Svetlana. *The Art of Describing: Dutch Art in the Seventeenth Century*. Chicago: University of Chicago Press, 1983.
Anderson, Frank J. *An Illustrated History of the Herbals*. New York: Columbia University Press, 1977.
Arnold, David. *The Problem of Nature: Environment, Culture and European Expansion*. London: Blackwell, 1996.
Audoin-Rouzeau, Frédérique. "Elevage et alimentation dans l'espace européen au Moyen Âge: Cartographie des ossements animaux." In *Milieux naturels, espaces sociaux: Études offertes à Robert Delort*. Edited by Elisabeth Mornet and Franco Morenzoni, 143–59. Paris: Publications de la Sorbonne, 1997.
Augustine. *The City of God against the Pagans*. Edited and translated by R.W. Dyson. Cambridge: Cambridge University Press, 1998.
———, Bishop of Hippo. *The City of God against the Pagans*. Translated by George E. McCracken et al. 7 vols. Cambridge, Mass.: Harvard University Press, 1957–72.
Bailey, Mark. "*Per impetum Maris*, Natural Disaster and Economic Decline in Eastern England, 1275–1350." In *Before the Black Death: Studies in the "Crisis" of the Early Fourteenth Century*. Edited by B.M.S. Campbell, 184–208. Manchester: Manchester University Press, 1991.

Baillie, M. G. L. "Dendrochronology Provides an Independent Background for Studies of the Human Past." In *L'uomo e la foresta secc. XIII–XVIII*. Edited by Simonetta Cavaciocchi. Instituto Internazionale di Storia Economica "F. Datini," Prato, Serie II—Atti delle "Settimane di Studi" e altri Convegni. Vol. 27, 99–119. Florence: Monnier, 1996.

———. "Putting Abrupt Environmental Change Back into Human History." In *Environments and Historical Change: The Linacre Lectures 1998*. Edited by Paul Slack, 46–75. Oxford: Oxford University Press, 1999.

Baker, Steve. *Picturing the Beast: Animals, Identity, and Representation*. 2nd ed. Champaign: University of Illinois Press, 2001.

———. "Sloughing the Human." In *Zoontologies: The Question of the Animal*. Edited by Cary Wolfe, 147–64. Minneapolis: University of Minneapolis Press, 2003.

Bale, Anthony. "Fictions of Judaism in England Before 1290." In *The Jews in Medieval Britain: Historical, Literary and Archaeological Perspectives*. Edited by Patricia Skinner, 129–44. Woodbridge: Boydell Press, 2003.

Balée, William. "Ecology, Historical." In *Encyclopedia of World Environmental History*. Vol. 1. Edited by Shepard Krech III, John R. McNeill, and Carolyn Merchant, 392–96. New York and London: Routledge, 2004.

Balibar, Étienne, and Immanuel Wallerstein. *Race, Nation, Class: Ambiguous Identities*. Translated by Chris Turner. London: Verso, 1991.

Barlaeus, Caspar. *Rerum per octennium in Brasilia et alibi nuper gestarum sub. Praefectura illustrissimi Comitis I. Mavritii, Nassoviae & Comitis, nunc Vesaliae Gubernatoris & Equitatus Faederatorum Belgii Ordd. Sub Avriaco Ductoris Historia*. Amsterdam: Johan Blaeu, 1647.

Barlow, L. K., J. C. Rogers, M. C. Serreze, and R. G. Barry. "Aspects of Climate Variability in the North Atlantic Sector: Discussion and Relation to the Greenland Ice Sheet Project 2 High-resolution Isotopic Signal." *Journal of Geophysical Research* 102 (C12):26 (1997): 333–44.

Bartholomaeus Anglicus. *De rerum Proprietatibus*. Frankofurti: apud Wolfgangum Richterum, 1601. Facsimile reprint, Frankfurt: Minerva, 1964.

Bartlett, Robert. *England under the Norman and Angevin Kings, 1075–1225*. Oxford: Clarendon Press, 2000.

Bartosiewicz, László. "People and Animals: The Archaeozoologist's Perspective." In *People and Nature in Historical Perspective*. Edited by József Laszlovszky and Péter Szabó, 23–34. Budapest: Central European University, Department of Medieval Studies, and Archaeolingua, 2003.

Baxter, Ron. *Bestiaries and Their Users in the Middle Ages*. Phoenix Mill: Sutton, 1998.

Beltran, E. "Jacques Legrand OESA: Sa vie et son oeuvre." *Augustiniana* 24 (1974): 132–60, 387–414.

Bennett, Joan. *Sir Thomas Browne: "A Man of Achievement in Literature."* Cambridge: Cambridge University Press, 1962.

Bergvelt, Ellinoor, and Renée Kistemaker, eds. *De wereld binnen handbereik: Nederlandse kunst- en rariteitenverzamelingen, 1585–1735.* Zwolle: Waanders Uitgevers, 1992.

Bernard Silvestris. *The Cosmographia of Bernardus Silvestris.* Translated by Winthrop Wetherbee. New York: Columbia University Press, 1973.

———. *Bernardus Silvestris: Cosmographia.* Edited by Peter Dronke. Leiden: Brill, 1978.

Biard, Joel. "Le statut du movement dans la philosophie naturelle buridanienne." In *La nouvelle physique du XIVe siècle.* Edited by Stefano Caroti and Pierre Souffrin, 141–59. Florence: Olschki, 1997.

Biddick, Kathleen. *The Other Economy: Pastoral Husbandry on a Medieval Estate.* Berkeley and Los Angeles: University of California Press, 1989.

Bildhauer, Bettina, and Robert Mills, eds. *The Monstrous Middle Ages.* Toronto: University of Toronto Press, 2003.

Biringuccio, Vannoccio. *Pirotechnia* (1540). Translated by Cyril Stanley Smith and Martha Teach Gnudi. Cambridge, Mass.: Harvard University Press, 1966.

Blockmans, Wim, and Walter Prevenier. *The Promised Lands: The Low Countries under Burgundian Rule, 1369–1530.* Translated by Elizabeth Fackelman. Philadelphia: University of Pennsylvania Press, 1999.

Boehrer, Bruce. *Shakespeare Among the Animals: Nature and Society in the Drama of Early Modern England.* Basingstoke: Palgrave, 2002.

Boesch, Hans, ed. "Urkunden und Auszüge aus dem Archiv und der Bibliothek des Germanischen Museums in Nürnberg." *Jahrbuch der kunsthistorischen Sammlungen der allerhoechsten Kaiserhauses* 7, Wien 1888.

Bordessoule, Nadine. *De proies et d'ombres: Escritures de la chasse dans la littérature française du XIVe siècle.* New York: Peter Lang, 2000.

Bork, Hans-Rudolf, and Gabrielle Schmidtchen. "Boden: Entwicklung, Zerstörung und Schutzbedarf in Deutschland." *Geographische Rundschau* 53:5 (May 2001): 4–9.

Boswell, John. *Christianity, Social Tolerance, and Homosexuality: Gay People in Western Europe from the Beginning of the Christian Era to the Fourteenth Century.* Chicago: University of Chicago Press, 1980.

Botkin, Daniel. *Discordant Harmonies: A New Ecology for the Twenty-First Century.* New York: Oxford University Press, 1990.

Bott, Gerhard, ed. *Wenzel Jamnitzer und die Nürnberger Goldschmiedekunst 1500–1700.* Catalog of the Germanisches Nationalmuseum, Nuremberg. Munich: Klinkhardt & Bierman, 1985.

Bourin-Derruau, Monique, Daniel Le Blévec, Claude Raynaud, and Laurent Schneider. "Le littoral languedocien au Moyen Âge." In *Castrum 7: Zones côtières littorales dans le monde méditerranéen au Moyen Âge: Defense, peuplement, mise en valeur.* Actes du colloque international organisé par l'École française de Rome et la Casa de Velázquez, en collaboration avec le Collège de France et le Centre interuniversitaire d'histoire et d'archéologie médiévales. UMR 5648–Université Lyon II–C.N.R.S.–E.H.E.S.S, Rome, 23–26 octobre 1996. Edited by Jean-Marie Martin, 345–423. Rome-Madrid: École française de Rome, Casa de Velázquez, 2001.

Boxer, Charles Ralph. *The Dutch in Brazil.* Oxford: Clarendon Press, 1957.

Brasil-Holandês/Dutch Brazil. 17 vols. Rio de Janeiro: Editora Index; and Petropolis: Editora Index, 1995–2002.

Brázdil, Rudolf. "Patterns of Climate in Central Europe since Viking Times." In *Climate Development and History of the North Atlantic Realm.* Edited by G. Wefer et al., 355–68. Berlin and Heidelberg: Springer Verlag, 2002.

———. "Historical Climatology and Its Progress after 1990." In *People and Nature in Historical Perspective.* Edited by József Laszlovszky and Péter Szabó, 197–228. Budapest: Central European University, Department of Medieval Studies, and Archaeolingua, 2003.

Bredekamp, Horst. *The Lure of Antiquity and the Cult of the Machine.* Translated by Allison Brown. Princeton: Markus Wiener, 1995.

Bresc, Henri. "La pêche dans l'espace économique normand." In *Terra e uomini nel Mezzogiorno normanno-svevo.* Atti delle settime giornate normanno-sveve, Bari 15–17 ottobre 1985. Edited by G. Musca, 271–329. Bari: Centro di studi normanno-svevi Università degli Studi di Bari, 1987.

———. "Pêche et habitat en méditerranée occidentale." In *Castrum 7: Zones côtières littorales dans le monde méditerranéen au Moyen Âge: Defense, peuplement, mise en valeur.* Actes du colloque international organisé par l'École française de Rome et la Casa de Velázquez, en collaboration avec le Collège de France et le Centre interuniversitaire d'histoire et d'archéologie médiévales. UMR 5648–Université Lyon II–C.N.R.S.–E.H.E.S.S, Rome, 23–26 octobre 1996. Edited by Jean-Marie Martin, 525–39. Rome-Madrid: École française de Rome, Casa de Velázquez, 2001.

Bridgwater, Benjamin, and John Dunton. *Religio Bibliopolae: In Imitation of Dr. Browns Religio Medici With a Supplement to It.* London, 1691.

Brienen, Rebecca Parker. "Albert Eckhout and Frans Post: Two Dutch Artists in Colonial Brazil." In *Brazil: Body and Soul.* Edited by Edward J. Sullivan, 62–99. New York: Guggenheim Museum, 2001.

———. "Albert Eckhout's Paintings and the Vrijburg Palace in Dutch Brazil." In *Albert Eckhout volta ao Brasil / Albert Eckhout returns to Brazil, 1644–2002,* 81–91. Copenhagen: Nationalmusseet, 2002.

———. "Albert Eckhout's Paintings of the '*wilde natien*' of Brazil and Africa." In *Picturing the Exotic, 1550–1950: Peasants and Outlandish Peoples in Nether-*

landish Art, 107–31. *Nederlands Kunsthistorisch Jaarboek / Netherlands Yearbook for the History of Art* 53, 2002.

———. *Visions of Savage Paradise: Albert Eckhout, Court Painter in Colonial Dutch Brazil.* Amsterdam: Amsterdam University Press, 2006.

Brown, Neville. *History and Climatic Change: A Eurocentric Perspective.* New York: Routledge, 2001.

Browne, Thomas. *Pseudodoxia Epidemica.* Edited by Robin Robbins. Oxford: Oxford University Press, 1981.

———. *Religio Medici.* In *The Major Works.* Edited by C. A. Patrides. Harmondsworth: Penguin, 1977.

Buckland, R. C., et al. "Bioarchaeological and Climatological Evidence for the Fate of the Norse Farmers in Medieval Greenland." *Antiquity* 70 (1996): 88–96.

Bullough, Vern, and Cameron Campbell. "Female Longevity and Diet in the Later Middle Ages." *Speculum* 55 (1980): 317–25.

Buridan, John. *Iohannis Buridani Quaestiones super libris quattuor de caelo et mundo.* Edited by Ernest A. Moody. Cambridge, Mass.: The Medieval Academy of America, 1942.

———. *Acutissimi philosophi reverendi Magistri Johannis Buridani subtillissime questiones super octo phisicorum libros Aristotelis* . . . Paris, 1509. Reprint, Frankfurt: Minerva, 1964.

———. *Les Questiones super tres libros Metheorum Aristotelis de Jean Buridan: Étude suive de l'édition du livre I.* Edited by Sylvia Bages. Thèse de l'Ecole de Chartes, 1986.

———. *Joannis Buridani Expositio et Quaestiones in Aristotelis De caelo.* Edited by Benoît Patar. Louvain: Éditions Peeters, 1996.

Butters, Suzanne B. *The Triumph of Vulcan: Sculptors' Tools, Porphyry, and the Prince in Ducal Florence.* 2 vols. Florence: Leo S. Olschki, 1996.

Buvelot, Quentin, ed. *Albert Eckhout: A Dutch Artist in Brazil.* Ex. cat. Mauritshuis, The Hague. Zwolle: Waanders Publishers, 2004.

Campbell, Bruce M. S., and Mark Overton. "A New Perspective on Medieval and Early Modern Agriculture: Six Centuries of Norfolk Farming, c.1250–c.1850." *Past & Present* 141 (1993): 38–105.

Capon, Brian. *Botany for Gardeners.* Rev. ed. Portland, Ore.: Timber Press, 2005.

Carmody, Francis, ed. "Physiologus Latinus, versio Y." *University of California Publications in Classical Philology* 12 (1941): 95–134.

Carson, Rachel. *Silent Spring.* Boston: Houghton-Mifflin, 1962.

Cassiodorus. *Variarum libri duodecim.* Edited by A. J. Fridh. Corpus Christianorum, Series Latina, vol. 96. Turnhout: Brepols, 1973.

———. *Cassiodorus: Variae.* Edited and translated by S. J. B. Barnish. Liverpool and Philadelphia: Liverpool University Press, 1992.

La Chace dou Cerf [ca. 1290]. Edited and translated by Gunnar Tilander. Cynegetica 7. Stockholm: Offset-Lito, 1960.

La Chasse, Les Dits du bon chien Souillard, et Les Louanges de Madame Anne de France. Edited by Gunnar Tilander. Cynegetica 6. Lund: Bloms, 1959.

Chaucer, Geoffrey. *Works.* In *The Riverside Chaucer.* Edited by Larry D. Benson. 3rd ed. Boston: Houghton Mifflin, 1987.

Chrétien de Troyes. *The Story of the Grail (Li Contes del Graal or Perceval).* Edited by Rupert Pickens. Translated by William W. Kibler. New York: Garland, 1990.

Chuine, Isabelle, Pascal Yiou, Nicolas Viovy, Bernard Seguin, Valérie Daux, and Emmanuel Le Roy Ladurie. "Historical Phenology: Grape Ripening As a Past Climate Indicator." *Nature* 432 (18 November 2004): 289–90.

Clagett, Marshall. *The Science of Mechanics in the Middle Ages.* Madison: University of Wisconsin Press, 1959.

———. "Nicole Oresme and Medieval Scientific Thought." *Proceedings of the American Philosophical Society* 108 (1964): 298–310.

———. *Archimedes in the Middle Ages.* 3 vols. Philadelphia: American Philosophical Society, 1964–76.

Clark, Willene B. *A Medieval Book of Beasts: The Second Family Bestiary.* Woodbridge: Boydell Press, 2006.

Clavel, Benoît. *L'Animal dans l'alimentation médiévale et moderne en France du nord (XIIe–XVIIe siècles).* Revue Archéologique de Picardie, No. Spécial 19. N.p.: CRAVO, 2001.

Cohen, Jeffrey Jerome. *Medieval Identity Machines.* Minneapolis: University of Minnesota Press, 2003.

———. *Hybridity, Identity and Monstrosity in Medieval Britain: On Difficult Middles.* New York: Palgrave Macmillan, 2006.

Cole, F.J. *Early Theories of Sexual Generation.* Oxford: Oxford University Press, 1930.

Collingwood, R.G. *The Idea of Nature.* Oxford: Clarendon Press, 1945.

Collins, Minta. *Medieval Herbals: The Illustrative Tradition.* Toronto: University of Toronto Press, 2000.

Comet, Georges. "Les céréales du Bas-Empire au Moyen Age." In *The Making of Feudal Agricultures?* Edited by Miquel Barceló and François Siguat, 131–76. Leiden: Brill, 2004.

Cooter, William S. "Ecological Dimensions of Medieval Agrarian Systems." *Agricultural History* 52 (1978): 458–77.

———. "Environmental, Ecological, and Agricultural Systems: Approaches to Simulation Modeling Applications for Medieval Temperate Europe." In *Archaeological Approaches to Medieval Europe.* Edited by Kathleen Biddick, 159–70. Kalamazoo, Mich.: Medieval Institute Publications, 1984.

Corrêa do Lago, Bia, ed. *Frans Post e o Brasil Holandês na coleção do Instituto Ricardo Brennand / Frans Post and Dutch Brazil in the Collection of Instituto*

Ricardo Brennand. Ex. cat. Instituto Ricardo Brennand. Recife: Ricardo Brennand, 2003.
Courtenay, William. "The Early Career of Nicole Oresme." *Isis* 91 (2000): 542–48.
Crawford, Patricia. "Sexual Knowledge in England, 1500–1750." In *Sexual Knowledge, Sexual Science: The History of Attitudes to Sexuality.* Edited by Roy Porter and Mikuláš Teich, 82–106. Cambridge: Cambridge University Press, 1994.
Creager, Angela N. H., and William Chester Jordan, eds. *The Animal/Human Boundary: Historical Perspectives.* Rochester: Rochester University Press, 2002.
Crook, D. S., D. J. Siddle, J. A. Dearing, and R. Thompson. "Human Impact on the Environment in the Annecy Petit Lac Catchment, Haute Savoie: A Documentary Approach." *Environment and History* 10 (2004): 247–84.
Crosby, A. W. *Ecological Imperialism: The Biological Expansion of Europe, 900–1900.* Cambridge: Cambridge University Press, 1986.
Crouch, David. *The Normans: The History of a Dynasty.* London: Hambledon and London, 2002.
Crumley, Carole L., ed. *Historical Ecology: Cultural Knowledge and Changing Landscapes.* Santa Fe: School of American Research Press, 1994.
Cummins, John. *The Hound and the Hawk: The Art of Medieval Hunting.* New York: St. Martin's, 1988.
———. "*Veneurs s'en vont en Paradis:* Medieval Hunting and the 'Natural' Landscape." In *Inventing Medieval Landscapes: Senses of Place in Western Europe.* Edited by John Howe and Michael Wolfe, 33–56. Gainesville: University Press of Florida, 2002.
Curley, Michael J. *Physiologus.* Austin: University of Texas Press, 1979.
Dalché, Patrick Gautier. "L'influence de Jean Buridan: L'habitabilité de la terre selon Dominicus de Clavisio." In *Comprendre et maîtriser la nature au Moyen Age: Mélanges d'histoire des sciences offerts à Guy Beaujouan,* 101–15. Geneva: Librairie Droz, 1994.
Daston, Lorraine, and Fernando Vidal. "Introduction: Doing What Comes Naturally." In *The Moral Authority of Nature,* 1–20. Chicago: University of Chicago Press, 2004.
Davis, R. H. C. *The Medieval Warhorse: Origin, Development and Redevelopment.* London: Thames and Hudson, 1989.
D'Avray, D. L. *The Preaching of the Friars: Sermons Diffused from Paris before 1300.* Oxford: Clarendon Press, 1985.
Dekkers, Midas. *Dearest Pet: On Bestiality.* Translated by Paul Vincent. London: Verso, 1994.
Delort, Robert, and Francois Walter. *Histoire de l'environnement européen.* Paris: Universitaires de France, 2001.

Derrida, Jacques. "The Animal That Therefore I Am (More to Follow)." Translated by David Wills. *Critical Inquiry* 28 (2002): 369–418.

———. "And Say the Animal Responded?" Translated by David Wills. In *Zoontologies: The Question of the Animal*. Edited by Cary Wolfe, 121–46. Minneapolis: University of Minnesota Press, 2003.

Dickinson, William R. "Changing Times: The Holocene Legacy." *Environmental History* 5 (2000): 483–502.

Digby, Sir Kenelm. *Observations upon "Religio Medici."* London, 1643.

———. *A Discourse Concerning the Vegetation of Plants*. London, 1661.

Dobson, R. B. *The Jews of Medieval York and the Massacre of March 1190*. University of York Borthwick Papers 45. York: University of York, 1974; rev. 1996.

Doering, Oscar. "Des Augsburger Patriciers Philipp Hainhofer Beziehungen zum Herzog Philipp II von Pommern-Stettin: Correspondenzen aus den Jahren 1610–1619." In *Quellenschriften für Kunstgeschichte und Kunsttechnik des Mittelalters und der Neuzeit*. NF Bd. 6. Vienna: Carl Graeser, 1894.

Douglas, Mary. *Purity and Danger: An Analysis of the Concepts of Pollution and Taboo*. London: Routledge, 1966.

Du Cange, Charles du Fresne. *Glossarium mediae et infimae Latinitas*. Niort: L. Fabre, 1883–87.

Duhem, Pierre. *Le système du monde*. Paris: Hermann, 1958.

Dunin-Wąsowicz, Teresa. "Environnement et habitat: La rupture d'équilibre du XIIIe siècle dans la Grande Plaine Européene." *Annales E.S.C.* 35:5 (Septembre–Octobre 1980): 1026–45.

———. "Natural Environment and Human Settlement over the Central European Lowland in the 13th Century." In *Silent Countdown: Essays in European Environmental History*. Edited by Peter Brimblecombe and Christian Pfister, 92–105. Berlin, Heidelberg, New York: Springer Verlag, 1990.

Dyer, Christopher. "English Diet in the Later Middle Ages." In *Social Relations and Ideas*. Edited by T. H. Aston et al., 191–216. Cambridge: Cambridge University Press, 1983.

———. "Changes in Diet in the Late Middle Ages: The Case of Harvest Workers." *The Agricultural History Review* 36 (1988): 21–38.

Economou, George D. *The Goddess Natura in Medieval Literature*. 1972; reprint, Notre Dame, Ind.: University of Notre Dame Press, 2002.

Effmert, Viola. ". . . ein schön kunstlich silbre vergult truhelein . . . Wenzel Jamnitzers Prunkkassette in Madrid." *Anzeiger des Germanischen Nationalmuseums* (1989): 131–58.

Egbert, Virginia Wylie. *On the Bridges of Mediaeval Paris: A Record of Early Fourteenth-Century Life*. Princeton: Princeton University Press, 1974.

Egmond, Florike. "Clusius, Cluyt, Saint Omer: The Origins of the Sixteenth-Century Botanical and Zoological Watercolours in Libri Picturati A. 16–30." *Nuncius* 20 (2005): 11–67.

Egmond, Florike, Paul G. Hoftijzer, and Robert P.W. Visser, eds. *Carolus Clusius in a New Context: Towards a Cultural History of a Renaissance Naturalist*. Amsterdam: Edita Publishing House of the Royal Dutch Academy, 2007.

Eichberger, Dagmar. *Leben mit Kunst: Wirken durch Kunst. Sammelwesen und Hofkunst unter Margarete von Österreich, Regentin der Niederlande*. Turnhout: Brepols, 2002.

Eiriksson, J., K. I. Knudsen, H. Haflidason, and J. Heinemeier. "Chronology of Late Holocene Climatic Events in the Northern North Atlantic Based on AMS C-14 Dates and Tephra Markers from the Volcano Hekla, Iceland." *Journal of Quaternary Science* 15 (2000): 573–80.

Emanuelsson, U. "The Relationship of Different Agricultural Systems to the Forest and Woodlands of Europe." In *Human Influence on Forest Ecosystems Development in Europe*. Edited by F. Salbitano, 169–78. Bologna: [n.p.], 1989.

Ervynck, Anton. "Medieval Castles as Top-Predators of the Feudal System: An Archaeozoological Approach." *Chateau Gaillard* 15 (1992): 151–59.

———. "Following the Rule? Fish and Meat Consumption in Monastic Communities in Flanders (Belgium)." In *Environment and Subsistence in Medieval Europe*. Edited by Guy De Boe and Frans Verhaeghe, 67–81. Papers of the "Medieval Europe Brugge 1997" Conference, vol. 9. I.A.P. Rapporten, vol. 9. Zellik, Belgium: Instituut voor het Archeologisch Patrimonium, 1997.

———. "*Orant, pugnant, laborant*: The Diet of the Three Orders in the Feudal Society of Medieval North-western Europe." In *Behaviour behind Bones: The Zooarchaeology of Ritual, Religion, Status and Identity*. Edited by Sharyn Jones O'Day, Wim van Neer, and Anton Ervynck, 215–23. Proceedings of the 9th Conference of the International Council of Archaeozoology, Durham, August 2002. Oxford: Oxbow Books, 2004.

Evelyn, John. *Publick Employment and an Active Life*. London, 1667.

———. *Sylva, or a Discourse of Forest Trees*. London: Doubleday, 1908.

The Exeter Book. Edited by George P. Krapp and Elliott V. K. Dobbie. New York: Columbia University Press, 1936.

Farley, John. *The Spontaneous Generation Controversy from Descartes to Oparin*. Baltimore: Johns Hopkins University Press, 1977.

Faust, D., C. Zielhofer, F. Diaz del Olmo, and R. B. Escudero. "Fluvial Record of Late Pleistocene and Holocene Geomorphic Change in Northern Tunisia—Global, Regional or Local Climatic Causes?" *PAGES Past Global Changes News* 13:1 (April 2005): 13–14.

Ferguson, Charles A. "Absence of Copula and the Notion of Simplicity: A Study of Normal Speech, Baby Talk, Foreigner Talk, and Pidgins." In *Pidginization and Creolization of Languages*. Edited by Dell Hymes, 141–50. Cambridge: Cambridge University Press, 1971.

———. "Baby Talk in Six Languages." In *Language Structure and Language Use: Essays by Charles A. Ferguson*. Edited by Anwar S. Dil, 113–33. Stanford, Calif.: Stanford University Press, 1971.

Filotas, Bernadette. *Pagan Survivals, Superstitions and Popular Cultures in Early Medieval Pastoral Literature*. Toronto: Pontifical Institute of Mediaeval Studies, 2005.

Filuk, J. "Biologiczno-rybacka charakterystyka ichtiofauny zalewu wiślanego na tle badań paleoichtiologicznych, historycznych i wspólczesnych." *Pomorania antiqua* 2 (1968): 146–48.

Findlen, Paula. *Possessing Nature*. Berkeley and Los Angeles: University of California Press, 1994.

Finke, Laurie A., and Martin B. Shichtman. *King Arthur and the Myth of History*. Gainesville: University Press of Florida, 2004.

Fischer-Kowalski, Marina. "Ecology, Social." In *Encyclopedia of World Environmental History*. Vol. 1. Edited by Shepard Krech III, John R. McNeill, and Carolyn Merchant, 396–400. New York and London: Routledge, 2004.

Fish, Stanley E. *Self-Consuming Artifacts: The Experience of Seventeenth-Century Literature*. Berkeley: University of California Press, 1972.

Flores, Nona C. *Animals in the Middle Ages*. New York: Garland, 2000.

Fox, H. S. A. "Some Ecological Dimensions of Medieval Field Systems." In *Archaeological Approaches to Medieval Europe*. Edited by Kathleen Biddick, 119–58. Kalamazoo, Mich.: Medieval Institute Publications, 1984.

Freedberg, David. *The Eye of the Lynx: Galileo, His Friends, and the Beginnings of Modern Natural History*. Chicago: University of Chicago Press, 2002.

Frenzel, Burkhard, ed. *Climatic Trends and Anomalies in Europe 1675–1715: High Resolution Spatio-Temporal Reconstructions from Direct Meteorological Observations and Proxy Data: Methods and Results*. Stuttgart and New York: G. Fischer, 1994.

Fuciková, Eliska, ed. *Prag um 1600: Beiträge zur Kunst und Kultur am Hofe Rudolfs II*. Freren and Emsland: Luca, 1988.

———, et al., eds. *Rudolf II and Prague: The Court and the City*. London: Thames and Hudson, 1997.

Fudge, Erica. *Perceiving Animals: Humans and Beasts in Early Modern English Culture*. Basingstoke: Palgrave, 2000.

———. *Brutal Reasoning: Animals, Rationality, and Humanity in Early Modern England*. Ithaca, N.Y.: Cornell University Press, 2006.

Gál, Erika. "Adaptation of Different Bird Species to Human Environments." In *People and Nature in Historical Perspective*. Edited by József Laszlovszky and Péter Szabó, 121–38. Budapest: Central European University, Department of Medieval Studies, and Archaeolingua, 2003.

Gasking, Elizabeth B. *Investigations into Generation, 1651–1828*. Baltimore: Johns Hopkins University Press, 1967.

Gaston Phébus. *Gaston Phébus: Livre de chasse*. Edited by Gunnar Tilander. Cynegetica 18. Karlshamn: Johanssons, 1971.
Gauricus, Pomponius. *De Sculptura*, ca. 1503. Translated by Heinrich Brockhaus. Leipzig: F.A. Brockhaus, 1886.
Geertz, Clifford. "Deep Play: Notes on the Balinese Cockfight." In *The Interpretation of Cultures*, 412–53. New York: Basic Books, 1973.
Geoffrey of Monmouth. *History of the Kings of Britain*. Translated by Lewis Thorpe. London: Penguin, 1966.
———. *The Historia Regum Brittanie*. Vol 1. *Bern, Bürgerbibliothek MS 568 (the 'Vulgate' Version)*. Edited by Neil Wright. Cambridge: D.S. Brewer, 1984.
George, Wilma, and Brunsdon Yapp. *The Naming of the Beasts: Natural History in the Medieval Bestiary*. London: Duckworth, 1991.
Gesner, Konrad. *Historiae Animalium*. 5 vols. Zurich: C. Froschoverum, 1551–58.
Gesta Stephani. Edited and translated by K.R. Potter. Introduction and notes by R.H.C. Davis. Oxford: Oxford University Press, 1976.
Gildas. *De Excidio Britonum*. Edited and translated by Michael Winterbottom. London and Chichester: Phillimore, 1978.
Glaser, Rüdiger. *Klimageschichte Mitteleuropas. 1000 Jahre Wetter, Klima, Katastrophen*. Darmstadt: Wissenschaftliche Buchgesellschaft, 2001.
Glass, Bentley. "The Germination of the Idea of Biological Species." In *Forerunners of Darwin: 1745–1859*. Edited by Bentley Glass et al., 30–48. Baltimore: Johns Hopkins University Press, 1959.
Gleitman, Lila R., Elissa L. Newport, and Henry Gleitman. "The Current Status of the Motherese Hypothesis." *Journal of Child Language* 11 (1984): 43–79.
Grabmeyer, Johannes. *Europa im späten Mittelalter, 1250–1500: Eine Kultur- und Mentalitätsgeschichte*. Darmstadt: Wissenschaftliche Buchgesellschaft, 2004.
Grant, Edward, ed. *A Source Book in Medieval Science*. Cambridge, Mass.: Harvard University Press, 1974.
———. *The Foundations of Modern Science in the Middle Ages: Their Religious, Institutional, and Intellectual Contexts*. Cambridge: Cambridge University Press, 1996.
———. *God and Reason in the Middle Ages*. Cambridge: Cambridge University Press, 2001.
Green, S.W. "The Agricultural Colonization of Temperate Forest Habitats: An Ecological Model." In *The Frontier: Comparative Studies*. Vol. 2. Edited by W.W. Savage, Jr., and S. Thompson, 69–103. Norman: University of Oklahoma Press, 1979.
Grieco, Allen J. "Food and Social Classes in Late Medieval and Renaissance Italy." In *Food: A Culinary History from Antiquity to the Present*. Edited by Jean-Louis Flandrin and Massimo Montanari with Albert Sonnenfeld. Translated by Clarissa Botsford et al., 302–12. New York: Columbia University Press, 1999.

Gries, Christian. "Erzherzog Ferdinand II von Tirol und die Sammlungen auf Schloß Ambras." *Frühneuzeit-Info* 5 (1994): 7–37.
Grote, Andreas, ed. *Macrocosmos in Microcosmo: die Welt in der Stube. Zur Geschichte des Sammelns 1450 bis 1800*. Opladen: Leske & Budrich, 1994.
Grove, A. T., and Oliver Rackham. *The Nature of Mediterranean Europe: An Ecological History*. New Haven: Yale University Press, 2001.
Grove, Jean M. *The Little Ice Age*. London: Methuen, 1988.
———. "The Century Time-Scale." In *Time-Scales and Environmental Change*. Edited by T.S. Driver and G.P. Chapman, 39–87. New York: Routledge, 1996.
———. "The Onset of the Little Ice Age." In *History and Climate: Memories of the Future?* Edited by P.D. Jones, A.E.J. Ogilvie, T.D. Davies, and K.R. Briffa, 153–86. New York: Kluwer, 2001.
———. "The Initiation of the Little Ice Age in Regions round the North Atlantic." *Climatic Change* 48 (2001): 53–82.
Grove, Jean M., and R. Switsur. "Glacial Geological Evidence for the Medieval Warm Period." *Climatic Change* 30 (1994): 1–27.
Grove, Richard H. *Green Imperialism: Colonial Expansion, Tropical Island Edens and the Origin of Environmentalism, 1600–1800*. Cambridge: Cambridge University Press, 1995.
Grzimek, Bernard. *Grzimek's Encyclopedia of Mammals*. Edited by Sybil P. Parker. New York: McGraw-Hill, 1990.
Gudger, E.W. "George Marcgraf, the First Student of American Natural History." *The Popular Science Monthly* (September 1912): 250–74.
Guiot, Joël. "The Combination of Historical Documents and Biological Data in the Reconstruction of Climate Variations in Space and Time." In *European Climate Reconstructed from Documentary Data: Methods and Results*. Edited by Burkhard Frenzel, 94–104. Stuttgart: Fischer Verlag, 1992.
Gunn, J.D., ed. *The Years Without Summer: Tracing AD 536 and Its Aftermath*. BAR International Series 872. Oxford: Archaeopress, 2000.
Haberli, Wilfried, and Hanspeter Holzhauser. "Alpine Glacier Mass Changes during the Past Two Millennia." *PAGES Past Global Changes News* 11:1 (April 2003): 13–15.
Hanawalt, Barbara A. "Men's Games, King's Deer: Poaching in Medieval England." *Journal of Medieval and Renaissance Studies* 18 (1988): 175–93.
Haraway, Donna. *Companion Species Manifesto: Dogs, People, and Significant Otherness*. Chicago: Prickly Paradigm Press, 2003.
Hardouin de Fontaines-Guérin. *Le Trésor de vénerie*. [1394]. Edited by Jérôme Pichon. Paris: Techener, 1855.
Hardy, Alan, Anne Dodd, Graham D. Keevil, et al. *Ælfric's Abbey: Excavations at Eynsham Abbey, Oxfordshire, 1989–92*. Oxford Archaeology, Thames Valley Landscapes, 16, 402–6. Oxford: Oxford University School of Archaeology for Oxford Archaeology, 2003.

Harvey, William. *Disputations Touching the Generation of Animals* (1651). Translated by Gweneth Whitteridge. Oxford: Blackwell, 1981.
Hassig, Debra. *Medieval Bestiaries: Text, Image, Ideology.* Cambridge: Cambridge University Press, 1995.
Hastrup, Kirsten. *Culture and History in Medieval Iceland: An Anthropological Analysis of Structure and Change.* Oxford: Clarendon Press, 1985.
———. *Nature and Policy in Iceland 1400–1800: An Anthropological Analysis of History and Mentality.* Oxford: Clarendon Press, 1990.
Hauschke, Sven. "Wenzel Jamnitzer im Porträt: Der Künstler als Wissenschaftler." *Anzeiger des Germanischen Nationalmuseums* (2003): 127–36.
Havenstein, Daniela. *Democratizing Sir Thomas Browne: "Religio Medici" and Its Imitations.* Oxford: Oxford University Press, 1999.
Hayward, J. F. *Virtuoso Goldsmiths and the Triumph of Mannerism, 1540–1620.* London: Sotheby Park Bernet Publications, 1976.
———. "The Mannerist Goldsmiths: Wenzel Jamnitzer." *The Connoisseur* 164 (1976): 148–54.
Helfferich, Emil. *Prinz Johann Moritz von Nassau-Siegen und die Niederländischen Kolonien in Brasilien.* Berlin: Verlag Max Staercke, 1938.
Herlihy, David. "Attitudes toward the Environment in Medieval Society." In *Historical Ecology: Essays on Environment and Social Change.* Edited by Lester J. Bilsky, 100–116. Port Washington, N.Y.: Kennikat Press, 1980.
Highmore, Nathaniel. *The History of Generation.* London, 1651.
Hochstrasser, Julie Berger. "Visual Legacies of Slavery in Suriname? The Stakes of Not Seeing." In Proceedings of the conference "Globalization, Diaspora and Identity Formation," Paramaribo, Suriname, 2007.
Hoffmann, Richard C. *Land, Liberties, and Lordship in a Late Medieval Countryside: Agrarian Structures and Change in the Duchy of Wrocław.* Philadelphia: University of Pennsylvania Press, 1989.
———. "Frontier Foods for Late Medieval Consumers: Culture, Economy, Ecology." *Environment and History* 7 (2001): 131–67.
Honour, Hugh. *The New Golden Land: European Images of America from the Discoveries to the Present Time.* New York: Pantheon Books, 1975.
Hooykaas, Reijer. *Humanisme, Science et Reforme, Pierre de la Ramee.* Leyden: E. J. Brill, 1958.
Howe, James. "Fox Hunting as Ritual." *American Ethnologist* 8 (1981): 278–300.
Howell, James. *Epistolae Ho-Elianae.* London, 1645.
Huebert, Ronald. "The Private Opinions of Sir Thomas Browne." *Studies in English Literature, 1500–1900* 45 (2005): 117–34.
The Hunting Book of Gaston Phébus: Manuscrit français 616, Paris, Bibliothéque nationale. Introduction by Marcel Thomas and François Avril. Translated by Sarah Lane. Commentary by Wilhelm Schlag. London: Harvey Miller Publishers, 1998.

Huntley, Frank Livingstone. *Sir Thomas Browne: A Biographical and Critical Study.* Ann Arbor: University of Michigan Press, 1962.
Hüster-Plogmann, Heide, and André Rehazek. "1000 Years (6th to 16th Century) of Economic Life in the Heart of Europe: Common and Distinct Trends in Cattle Economy of the Baltic Sea Region and the Swiss Region of the Alpine Forelands." *Archaeofauna* 8 (1999): 123–33.
Impey, Oliver, and Arthur MacGregor, eds. *The Origins of Museums: The Cabinet of Curiosities in Sixteenth- and Seventeenth-Century Europe.* Oxford: Clarendon Press, 1985.
Jansen, Dirk Jacob. "Samuel Quicchebergs "Inscriptiones": de encyclopedische verzameling als hulpmiddel voor de wetenschap." In *Verzamelen: Van rariteitenkabinet tot kunstmuseum.* Edited by Ellinoor Bergvelt, Debora J. Meijers, and Mieke Rijnders, 55–76. Heerlen: Open Universiteit; Houten: Gaade, 1993.
John of Salisbury. *Letters of John of Salisbury.* Edited by W. J. Millor and H. E. Butler. Rev. C. N. L. Brooke. 2 vols. London: Thomas Nelson and Sons, 1955.
Jones, Eric. "The Bird Pests of British Agriculture in Recent Centuries." *The Agricultural History Review* 20 (1972): 107–25.
Julians Barnes Boke of Huntyng. Edited by Gunnar Tilander. Cynegetica 11. Karlshamn: Johanssons, 1964.
Kaufmann, Thomas DaCosta. *The Mastery of Nature: Aspects of Art, Science, and Humanism in the Renaissance.* Princeton: Princeton University Press, 1993.
Kaye, Joel. *Economy and Nature in the Fourteenth Century: Money, Market Exchange, and the Emergence of Scientific Thought.* Cambridge: Cambridge University Press, 1998.
Kemp, Martin. *The Science of Art: Optical Themes in Western Art from Brunelleschi to Seurat.* New Haven and London: Yale University Press, 1990.
———. "'Wrought by No Artist's Hand': The Natural, the Artificial, the Exotic, and the Scientific in Some Artifacts from the Renaissance." In *Reframing the Renaissance: Visual Culture in Europe and Latin America 1450–1650.* Edited by Claire Farago, 117–96. New Haven and London: Yale University Press, 1995.
Keynes, Geoffrey, ed. *The Works of Sir Thomas Browne.* London: Faber & Faber, 1931.
———, ed. *Religio Medici and Christian Morals.* New York: Thomas Nelson, 1940.
Keys, David. *Catastrophe.* London: Century, 1999.
Kiser, Lisa J. "Chaucer and the Politics of Nature." In *Beyond Nature Writing: Expanding the Boundaries of Ecocriticism.* Edited by Kathleen Wallace and Karla Armbruster, 41–59. Charlottesville: University Press of Virginia, 2001.
———. "Animals in Sacred Space: St. Francis and the Crib at Greccio." In *Speaking Images.* Edited by Charlotte Morse and R. F. Yeager, 56–73. Asheville, N.C.: Pegasus Press, 2001.

———. "The Garden of St. Francis: Plants, Landscape, and Economy in Thirteenth-Century Italy." *Environmental History* 8 (2003): 229–45.
———. "Animal Economies: The Lives of St. Francis in Their Medieval Contexts." *ISLE: Interdisciplinary Studies in Literature and the Environment* 11 (2004): 121–38.
———. "Attitudes towards Animals." In *Dictionary of the Middle Ages: Supplement I*, 17–21. New York: Charles Scribners, 2004.
Knottnerus, Otto S. "Malaria around the North Sea: A Survey." In *Climate Development and History of the North Atlantic Realm*. Edited by G. Wefer, W. Berger, K.-E. Wehre, and E. Jansen, 339–53. Berlin, Heidelberg, New York: Springer Verlag, 2002.
Kris, Ernst. "Der Stil 'Rustique': Die Verwendung des Naturabgusses bei Wenzel Jamnitzer und Bernard Palissy." *Jahrbuch der Kunsthistorischen Sammlungen in Wien*, NF 1 (1928): 137–207.
Kruger, Steven. "Conversion and Medieval Sexual, Religious, and Racial Categories." In *Constructing Medieval Sexuality*. Edited by Karma Lochrie, Peggy McCracken, and James A. Schultz, 158–79. Minneapolis: University of Minnesota Press, 1997.
Latour, Bruno. *The Politics of Nature: How to Bring the Sciences into Democracy*. Cambridge, Mass.: Harvard University Press, 2004.
L'Ecluse, Charles (Carolus Clusius). *Exoticorum Libri Decem: Quibus Animalium, Plantarum, Aromatum, aliorumque peregrinorum Fructuum historiae describuntur*. Antwerp: Ex Officina Plantiniana Raphelengii, 1605.
Leduff, Charlie. "At a Slaughterhouse, Some Things Never Die." In *Zoontologies: The Question of the Animal*. Edited by Cary Wolfe, 183–97. Minneapolis: University of Minnesota Press, 2003.
Leveau, Philippe. "L'archéologie des paysages et les époques historiques: Les grands aménagements agraires et leur signature dans le paysage (anthropisation des milieux et complexité des sociétés)." In *Milieux Naturels: Espaces Sociaux: Études offertes à Robert Delort*. Edited by Élisabeth Mornet and Franco Morenzoni, 71–84. Paris: Publications de la Sorbonne, 1997.
Lindberg, David C. *The Beginnings of Western Science: The European Scientific Tradition in Philosophical, Religious, and Institutional Context, 600 B.C. to A.D. 1450*. Chicago: University of Chicago Press, 1992.
Lingis, Alphonso. "Animal Body, Inhuman Face." In *Zoontologies: The Question of the Animal*. Edited by Cary Wolfe, 165–82. Minneapolis: University of Minnesota Press, 2003.
Les Livres du roy Modus et de la royne Ratio. Edited by Gunnar Tilander. 2 vols. Paris: Société des anciens texts français, 1932.
MacAloon, John J. "Olympic Games and the Theory of Spectacle in Modern Societies." In *Rite, Drama, Festival, Spectacle: Rehearsals toward a Theory of Cultural*

Performance. Edited by John J. MacAloon, 241–80. Philadelphia: Institute for the Study of Human Issues, 1984.

Makowiecki, Daniel. *Hodowla oraz użytkowanie zwierząt na Ostrowie Lednickim w średniowieczu: Studium archeozoologiczne*. Bibliotka studiów lednickich. Vol. 6. Poznań: Muzeum Pierwszych Piastów na Lednicy, 2001.

Marcgraf, George. *Brasilia qua parte paret Belgis*. Amsterdam: Johan Blaeu, 1657.

Marvin, Garry. "Unspeakability, Inedibility, and the Structures of Pursuit in the English Foxhunt." In *Representing Animals*. Edited by Nigel Rothfels, 139–58. Bloomington: Indiana University Press, 2002.

Mason, Peter. *Infelicities: Representations of the Exotic*. Baltimore: Johns Hopkins University Press, 1998.

The Master of Game by Edward, Second Duke of York. Edited by Wm. A. and F. Baillie-Grohman. London: Ballantyne, Hanson & Co., 1904.

The Master of Game by Edward, Second Duke of York. Translated by Wm. A. and F. Baillie-Grohman. London: Chatto and Windus, 1909. [English translation of above.]

Mauss, Marcel. *The Gift: The Form and Reason for Exchange in Archaic Societies*. Translated by W. D. Halls. New York: W. W. Norton, 2000.

McCulloch, Florence. *Medieval Latin and French Bestiaries*. Chapel Hill: University of North Carolina Press, 1960.

McGovern, Thomas H. "The Demise of Norse Greenland." In *Vikings: The North Atlantic Saga*. Edited by William W. Fitzhugh and Elizabeth Ward, 327–39. Washington, D.C.: Smithsonian Institution Press, 2000.

———, et al. "Northern Islands, Human Error, and Environmental Degradation: A View of Social and Ecological Change in the Medieval North Atlantic." *Human Ecology* 16 (1988): 225–70.

McNeill, John. "Observations on the Nature and Culture of Environmental History." *History and Theory* 42 (2003): 5–43.

Meadow, Mark A. "Merchants and Marvels: Hans Jacob Fugger and the Origins of the Wunderkammer." In *Merchants and Marvels: Commerce, Science, and Art in Early Modern Europe*. Edited by Pamela H. Smith and Paula Findlen, 182–200. New York: Routledge, 2002.

Meeuwse, Mark. "Allies and Subjects: The Legal and Political Status of the Tupi Indians in Dutch Brazil, 1630–1654." Paper presented at the workshop "The Dutch Golden Age and the World," Columbia University, March 2007.

Mello, José Antônio Gonsalves de. *Tempo dos Flamengos*. 4th ed. Recife: Instituto Ricardo Brennand, 2001.

Merton, Egon Stephen. *Science and Imagination in Sir Thomas Browne*. New York: King's Crown Press, 1949.

———. "The Botany of Sir Thomas Browne." *Isis* 47 (1956): 161–71.

Mitchell, W. J. T., ed. *Landscape and Power*. 2nd ed. Chicago: University of Chicago Press, 2002.

Montanari, Massimo. "From the Late Classical Period to the Early Middle Ages." In *Food: A Culinary History from Antiquity to the Present*. Edited by Jean-Louis Flandrin and Massimo Montanari, with Albert Sonnenfeld. Translated by Clarissa Botsford, 165–85. New York: Columbia University Press, 1999.

Moody, Ernest A. "John Buridan on the Habitability of the Earth." *Speculum* 16 (1941): 415–25.

Moore, Sally F., and Barbara G. Myerhoff. "Introduction: Secular Ritual: Forms and Meanings." In *Secular Ritual*. Edited by Sally F. Moore and Barbara G. Myerhoff, 3–24. Assen: Van Gorcum, 1977.

Morineau, Michel. "Cataclysmes et calamités naturelles aux Pays-Bas septentrionaux XIe–XVIIIe siècles: Le travaille de la planète et la rétorsion des hommes." In *Les catastrophes naturelles dans l'Europe médiévale et moderne*. Edited by B. Bennassar, 42–59. Actes des XVes Journées Internationales d'Histoire de l'Abbaye de Flaran, 10, 11, 12 Septembre 1993. Toulouse: Presses Universitaires du Mirail, 1996.

Muñiz, Arturo Morales, and Dolores Carmen Morales Muñiz. "¿De quién es este ciervo?: Algunas consideraciones en torno a la fauna cinegética de la España medieval." In *El medio natural en la España medieval: Actas del I Congreso sobre ecohistoria e historia medieval*. Edited by Julián Clemente Ramos, 383–406. Cáceres: Universidad de Extremadura, 2001.

Muratova, Xenia. "The Bestiaries: An Aspect of Medieval Patronage." In *Art and Patronage in the English Romanesque*. Edited by Sarah Macready and F. H. Thompson, 118–44. London: Society of Antiquaries, 1986.

Niavis, Paulus. *Iudicium iovis in valle amenitatis habitum ad quod mortalis homo a terra tractus propter montifodinas in monte niveo aliisque multis perfectas ac demum parricidii accusatus*. Leipzig: Martin Landsberg, ca. 1492/95.

———. *Iudicium Iovis oder Das Gericht der Götter über den Bergbau*. Edited and translated by Paul Krenkel. Freiberger Forschungshefte, Reihe Kultur und Technik, D3, 13–38. Berlin: Akademie, 1953.

Nordenfalk, Carl. "Hatred, Hunting, and Love: Three Themes Relative to Some Manuscripts of Jean sans Peur." In *Studies in Late Medieval and Renaissance Painting in Honor of Millard Meiss*. Edited by Irving Lavin and John Plummer. 2 vols. Vol. 1, pp. 324–41 (essay); vol. 2, p. 114 (plates for essay). New York: New York University Press, 1977.

Ogilvie, A. E. "Climatic Changes in Iceland A.D. ca. 865 to 1598." *Acta Archaeologica* 61 (1990): 233–51.

———, and Thomas H. McGovern. "Sagas and Science: Climate and Human Impacts in the North Atlantic." In *Vikings: The North Atlantic Saga*. Edited by William W. Fitzhugh and Elizabeth Ward, 385–93. Washington, D.C.: Smithsonian Institution Press, 2000.

Ogilvie, Brian W. *The Science of Describing: Natural History in Renaissance Europe*. Chicago: University of Chicago Press, 2006.

Oldfield, F., and R. L. Clark. "Environmental History—The Environmental Evidence." In *The Silent Countdown: Essays in European Environmental History.* Edited by Peter Brimblecombe and Christian Pfister, 152–55. Berlin, Heidelberg, New York: Springer Verlag, 1990.

Oresme, Nicole. *Commentary on Aristotle's* De caelo, II, 2. In *Le livre du ciel et du monde.* Edited and translated by Albert Menut and Alexander Denomy. Madison: University of Wisconsin Press, 1968.

Ortolani, Franco, and Silvana Pagliuca. *La variazioni climatiche storiche e la prevedibilità delle modificazioni relative all'effetto serra.* Roma: Asociazione Italiana Nucleare, 2001.

———. "Cyclical Climatic-Environmental Changes in the Mediterranean Area (2500 BP–Present Day)." *PAGES Past Global Changes News* 11:1 (April 2003): 15–17.

Pálsson, Gísli. "The Idea of Fish: Land and Sea in the Icelandic World View." In *Signifying Animals: Human Meaning in the Natural World.* Edited by R. Willis, 119–33. London: One World Archaeology, 1990.

Parry, M. L. "Secular Climatic Change and Marginal Land." *Transactions of the Institute of British Geographers* 64 (1975): 1–13.

Patrides, C. A. "'Above Atlas His Shoulders': An Introduction to Sir Thomas Browne." Introduction to *The Major Works* by Sir Thomas Browne. Edited by C. A. Patrides. Harmondsworth: Penguin, 1977.

Pearsall, Derek, and Elizabeth Salter. *Landscapes and Seasons of the Medieval World.* Toronto: University of Toronto Press, 1973.

Pearson, Kathy L. "Nutrition and the Early-Medieval Diet." *Speculum* 72 (1997): 1–32.

Pechstein, Klaus. "Der Merkelsche Tafelaufsatz von Wenzel Jamnitzer." *Mitteilungen des Vereins für Geschichte der Stadt Nürnberg* 61 (1974): 90–121.

———. "Der Goldschmied Wenzel Jamnitzer." In *Wenzel Jamnitzer und die Nürnberger Goldschmiedekunst 1500–1700.* Edited by Gerhard Bott. Catalog of the Germanisches Nationalmuseum, Nuremberg, 67–70. Munich: Klinkhardt & Bierman, 1985.

Perlim, J. *História das florestas: A importancia da Madeira no desenvolvimento da civilização.* Rio de Janeiro: Editora Imago, 1992.

Pfister, Christian. *Das Klima der Schweiz von 1525–1860 und seine Bedeutung in der Geschichte von Bevölkerung und Landwirtschaft.* 2nd ed. Bern: Verlag Paul Haupt, 1985.

———. "Variations in the Spring-Summer Climate of Central Europe from the High Middle Ages to 1850." In *Long and Short Term Variability of Climate.* Edited by H. Wanner and U. Siegenthaler, 57–82. Berlin, Heidelberg, New York: Springer Verlag, 1988.

———. *Wetternachhersage: 500 Jahre Klimavariationen und Naturkatastrophen (1496–1995).* Bern: Paul Haupt, 1999.

Piskorski, J. "The Historiography of the So-called 'East Colonisation' and the Current State of Research." In *The Man of Many Devices, Who Wandered Full Many Ways . . . : Festschrift in Honor of János M. Bak.* Edited by Balázs Nagy and Marcell Sebők, 654–67. Budapest: Central European University Press, 1999.

Piso, Willem. *Historia naturalis Brasiliae.* Leiden: Francis Hack; Amsterdam: Elsevier, 1648.

Pliny the Elder. *Natural History.* Translated by H. Rackham. 10 vols. Loeb Classical Library. Cambridge, Mass.: Harvard University Press, 1940.

Pomian, Krzysztof. *Collectors and Curiosities: Paris and Venice, 1500–1800.* Translated by Elizabeth Wiles-Portier. Cambridge: Polity Press, 1990.

Postles, David. "Cleaning the Medieval Arable." *The Agricultural History Review* 37 (1989): 130–43.

Prag um 1600: Kunst und Kultur am Hofe Rudolfs II. [Ex. cat.] Ausstellung, Kulturstiftung Ruhr, Villa Essen, 10.6–30.10, 1988.

Preston, Claire. *Thomas Browne and the Writing of Early Modern Science.* Cambridge: Cambridge University Press, 2005.

Pretty, J. "Sustainable Agriculture in the Middle Ages on the English Manor." *The Agricultural History Review* 38 (1990): 1–19.

Quiccheberg, Samuel [1529–67]. *Inscriptiones vel Tituli Theatri Amplissimi.* Ed. Harriet Roth. In *Der Anfang der Museumslehre in Deutschland: Das Traktat "Inscriptiones vel Tituli Theatri Amplissimi" von Samuel Quiccheberg.* Berlin: Akademie Verlag, 2000.

Rackham, Oliver. *The History of the Countryside.* London: Dent, 1986.

———. *Trees and Woodland in the British Landscape: The Complete History of Britain's Trees, Woods & Hedgerows.* 1976; rev. ed., London: Dent, 1990.

Remigereau, François. "Tristan 'maître de vénerie' dans la tradition anglaise et dans le roman de Thomas." *Romania* 58 (1932): 218–37.

Ribémont, Bernard. "Mais où est donc le centre de la terre." In *Terres médiévales.* Edited by Bernard Ribémont, 261–76. Paris: Editions Klincksieck, 1993.

Richards, John F. *The Unending Frontier: An Environmental History of the Early Modern World.* Berkeley: University of California Press, 2003.

Riera-Melis, Antoni. "Society, Food and Feudalism." In *Food: A Culinary History from Antiquity to the Present.* Edited by Jean-Louis Flandrin and Massimo Montanari, with Albert Sonnenfeld. Translated by Clarissa Botsford et al., 251–67. New York: Columbia University Press, 1999.

Roberts, N. *The Holocene: An Environmental History.* Oxford: Blackwell, 1989.

Rogers, John. *The Matter of Revolution: Science, Poetry, and Politics in the Age of Milton.* Ithaca, N.Y.: Cornell University Press, 1996.

Rooney, Anne. *Hunting in Middle English Literature.* Cambridge: Boydell Press, 1993.

Ross, Alexander. *Arcana Microcosmi.* London, 1651.

Rossi, Paolo. *Philosophy, Technology, and the Arts in the Early Modern Era.* Translated by Salvator Attanasio. New York: Harper and Row, 1970.
Rothfels, Nigel, ed. *Representing Animals.* Bloomington: Indiana University Press, 2002.
Royster, Francesca. "'Working Like a Dog': African Labor and Racing the Human-Animal Divide in Early Modern England." In *Writing Race Across the Atlantic World: Medieval to Modern.* Edited by Philip D. Beidler and Gary Taylor, 113–34. New York: Palgrave Macmillan, 2005.
Ruddiman, William F. "The Anthropogenic Era Began Thousands of Years Ago." *Climatic Change* 61 (2003): 261–93.
———. "Early Anthropogenic Overprints on Holocene Climate." *PAGES Past Global Changes News* 12:1 (April 2004): 18–19.
———. "How Did Humans First Alter Global Climate?" *Scientific American* (March 2005): 46–53.
Ruhland, Florian. "Schweinehaltung in und vor der Stadt." In *Nürnberg: Archäologie und Kulturgeschichte.* Edited by Birgit Friedel and Claudia Frieser, 319–25. Nürnberg: Verlag Dr. Faustus, 1999.
Russell, Emily W. B. *People and the Land Through Time: Linking Ecology and History.* New Haven: Yale University Press, 1997.
Saalfeld, Diedrich. "Der Boden als Objekt der Aneigung." In *Von der Angst zur Ausbeutung: Umwelterfahrung zwischen Mittelalter und Neuzeit.* Edited by Ernst Schubert and Bernd Herrmann. Frankfurt: Fischer Verlag, 1994.
Salisbury, Joyce E. *The Beast Within: Animals in the Middle Ages.* New York: Routledge, 1994.
Salter, David. *Holy and Noble Beasts: Encounters with Animals in Medieval Literature.* Cambridge: D.S. Brewer, 2001.
Scheicher, Elisabeth. *Die Kunstkammer (Schloss Ambras).* Innsbruck: Kunsthistorisches Museum, 1977.
———. *Die Kunst- und Wunderkammern der Habsburger.* Vienna: Molden, 1979.
Schmalkalden, Caspar. *Die wundersamen Reisen des Caspar Schmalkalden nach West- und Ostindien, 1642–1652.* Leipzig: Brockhaus, 1983.
———. *Mit Kompass und Kanonen: Abenteuerliche Reisen nach Brasilien und Fernost, 1642–1652.* Edited by Wolfgang Joost. Stuttgart: Thienemann, 2002.
Schmidt, Benjamin. "Inventing Exoticism: The Project of Dutch Geography and the Marketing of the World, circa 1700." In *Merchants and Marvels: Commerce, Science, and Art in Early Modern Europe.* Edited by Paula Findlen and Pamela H. Smith, 347–70. London, 2002.
Schnapper, Antoine. *Le Géant, La Licorne et la Tulipe: Collections et Collectionneurs dans le France du XVIIe Siècle.* Paris: Flammarion, 1988.
Schultz, Eva. "Notes on the History of Collecting and of Museums in the Light of Selected Literature of the Sixteenth to the Eighteenth Century." *Journal of History of Collections* 2 (1990): 205–18.

Schwarz-Zanetti, Werner, Christian Pfister, Gabriela Schwarz-Zanetti, and Hannes Schüle. "The EURO-CLIMHIST Data Base—A Tool for Reconstructing the Climate of Europe in the Pre-instrumental Period from High Resolution Proxy Data." In *European Climate Reconstructed from Documentary Data: Methods and Results*. Edited by Burkhard Frenzel, 193–210. Stuttgart: Fischer Verlag, 1992.

Sedgwick, Eve Kosofsky. *Touching Feeling: Affect, Pedagogy, Performativity*. Durham, N.C.: Duke University Press, 2003.

Sharrock, Robert. *The History of the Propagation and Improvement of Vegetables by the Concurrence of Art and Nature*. Oxford, 1672.

Shiel, R. S. "Improving Soil Productivity in the Pre-fertilizer Era." In *Land, Labour, and Livestock. Historical Studies in European Agricultural Productivity*. Edited by Bruce M. S. Campbell and Mark Overton, 51–77. Manchester: Manchester University Press, 1991.

Siewers, Alfred K. "Landscapes of Conversion: Guthlac's Mound and Grendel's Mere as Expressions of Anglo-Saxon Nation-Building." *Viator* 34 (2003): 1–39.

Silva, Leonardo Dantas. *Holandeses em Pernambuco, 1630–1654*. Recife: Instituto Ricardo Brennand, 2005.

Sir Gawain and the Green Knight. Edited by J. R. R. Tolkien and E. V. Gordon. Oxford: Clarendon, 1967.

Smith, Jeffrey Chipps. "The Artistic Patronage of Philip the Good, Duke of Burgundy (1419–1467)." Ph.D. Dissertation, Columbia University, 1979.

Smith, Pamela H. *The Body of the Artisan: Art and Experience in the Scientific Revolution*. Chicago: University of Chicago Press, 2004.

Sonnlechner, Christoph. "The Establishment of New Units of Production in Carolingian Times: Making Early Medieval Sources Relevant for Environmental History." *Viator* 35 (2004): 21–48.

Sorabji, Richard. *Animal Minds and Human Morals: The Origin of the Western Debate*. London: Duckworth, 1993.

Sörlin, Sverker, and Paul Warde. "The Problem of Environmental History: A Re-Reading of the Field," *Environmental History* 12 (2007): 107–30.

Spiegelman, Art. *Maus: A Survivor's Tale*. London: André Deutsch, 1987.

Staden, Hans. *Wahrhaftiger Historia und Beschreibung einer Landtschaft der Wilden Nacketen Grimmigen Menschenfresser Leuthen in der Newenwelt America*. 1557.

Stanford, Michael. "The Terrible Thresholds: Sir Thomas Browne on Sex and Death." *English Literary Renaissance* 18 (1988): 413–23.

Steel, Karl. "'Elles were beest lich to man': The Problems and Uses of Animal Likeness in *Sidrak and Bokkus*." *Exemplaria* (forthcoming).

Steinberg, Theodore. "Down to Earth: Nature, Agency, and Power in History." *American Historical Review* 107 (2002): 798–820.

———. *Down to Earth: Nature's Role in American History*. New York: Oxford University Press, 2002.
Sterry, Peter. *A Discourse of the Freedom of the Will*. 1675.
Stock, Brian. *Myth and Science in the Twelfth Century: A Study of Bernard Silvester*. Princeton: Princeton University Press, 1972.
Strathern, Marilyn. *After Nature: English Kinship in the Late Twentieth Century*. Cambridge: Cambridge University Press, 1992.
Strubel, Armand, and Chantal de Saulnier. *La Poétique de la chasse au Moyen Age: Les livres de chasse du XIVe siècle*. Paris: Presses Universitaires de France, 1994.
Swann, Marjorie. *Curiosities and Texts: The Culture of Collecting in Early Modern England*. Philadelphia: University of Pennsylvania Press, 2001.
Sylla, Edith. "Medieval Quantification of Qualities: The Merton School." *Archive for History of Exact Sciences* 8 (1971): 7–39.
———. "The Oxford Calculators." In *The Cambridge History of Later Medieval Philosophy*. Edited by Norman Kretzman, Anthony Kenny, and Jan Pinborg, 541–63. Cambridge: Cambridge University Press, 1982.
Taunay, A., ed. *Jorge Marcgraf: Historia natural do Brasil. Edição do Museo Paulista*. Sao Paolo: Imprensa Oficial do Estado, 1942.
Thiébaux, Marcelle. "The Mediaeval Chase." *Speculum* 42 (1967): 260–74.
———. "The Mouth of the Boar as a Symbol in Medieval Literature." *Romance Philology* 22 (1968–69): 281–99.
———. *The Stag of Love*. Ithaca, N.Y.: Cornell University Press, 1974.
Thiessen, Erik D., Emily A. Hill, and Jenny R. Saffran. "Infant-Directed Speech Facilitates Word Segmentation." *Infancy* 7 (2005): 53–71.
Thomas of Chobham. *Summa de arte praedicandi*. Edited by F. Morenzoni. Corpus Christianorum, continuatio medievalis 82. Turnhout: Brepols, 1988.
Thomas, Keith. *Man and the Natural World: Changing Attitudes in England, 1500–1800*. Harmondsworth: Penguin, 1983.
Thorndike, Lynn. *A History of Magic and Experimental Science*. 3 vols. New York: Columbia University Press, 1934.
Tilander, Gunnar. *Nouveaux essais d'étymologie cynégétique*. Cynegetica 4. Lund: Bloms, 1957.
———. *Mélanges d'étymologie cynégétique*. Cynegetica 5. Lund: Bloms, 1958.
The Tretyse off Huntyng. Edited by Anne Rooney. Scripta: Medieval and Renaissance Texts and Studies 19. Brussels: UFSAL, 1987.
Turner, F. *Beyond Geography: The Western Spirit against Wilderness*. New Brunswick, N.J.: Rutgers University Press, 1983.
Van den Boogaart, Ernst. "Infernal Allies: The Dutch West India Company and the Tarairiu 1631–1654." In *Johan Maurits van Nassau-Siegen 1604–1679: A Humanist Prince in Europe and Brazil: Essays on the Occasion of the Tercentenary*

of His Death. Edited by Ernst Van den Boogaart, 519–38. The Hague: Johan Maurits van Nassau Stichting, 1979.

Vanderjagt, Arjo, and Klaas van Berkel, eds. *The Book of Nature in Antiquity and the Middle Ages.* Leuven: Peeters, 2005.

———, eds. *The Book of Nature in Early Modern and Modern History.* Leuven: Peeters, 2006.

Vartanian, Aram. "Spontaneous Generation." In *Dictionary of the History of Ideas.* Edited by Philip P. Wiener, 307–12. New York: Charles Scribner's Sons, 1973.

La Vénerie de Twiti: Le plus ancien traité de chasse écrit en Angleterre; la version anglaise du meme traité et Craft of Venery. Edited by Gunnar Tilander. Cynegetica 2. Uppsala: Almquist & Wiksells, 1956.

Verhulst, Adriaan. *The Carolingian Economy.* Cambridge: Cambridge University Press, 2002.

Voisenet, Jacques. *Bêtes et hommes dans le monde medieval: Le bestiare des clercs du Ve au XIIe siècle.* Turnhout: Brepols, 2000.

von Schlosser, Julius. *Die Kunst- und Wunderkammern der Spätrenaissance.* 1908; reprint, 2 vols., Braunschweig: Klinkhardt und Bierman, 1978.

Waddington, Raymond B. "The Two Tables in *Religio Medici.*" In *Approaches to Sir Thomas Browne: The Ann Arbor Tercentenary Lectures and Essays.* Edited by C.A. Patrides, 81–99. Columbia: University of Missouri Press, 1982.

The Wanderer. Edited by R.F. Leslie. Manchester: Manchester University Press, 1966.

Watanabe-O'Kelly, Helen. *Court Culture in Dresden: From Renaissance to Baroque.* Hampshire and New York: Palgrave, 2002.

Weiss, Gail. *Body Images: Embodiment as Incorporeality.* New York: Routledge, 1998.

Westra, Haijo Jan. "Bernard Silvester." In *Dictionary of the Middle Ages.* Edited by Joseph R. Strayer. Vol. 2, 194–95. New York: Charles Scribner's Sons, 1983.

Whitehead, P.J.P., and M. Boeseman. *A Portrait of Dutch 17th Century Brazil: Animals, Plants and People by the Artists of Johan Maurits of Nassau.* Amsterdam, Oxford, New York: North Holland Publishing, 1989.

Wilding, Michael. "*Religio Medici* in the English Revolution." In *Dragon's Teeth: Literature in the English Revolution,* 89–113. Oxford: Oxford University Press, 1987.

William of Malmesbury. *Gesta Regum Anglorum: The History of the English Kings.* 2 vols. Edited and translated by R.A.B. Mynors, completed by R.M. Thomson and M. Winterbottom. Oxford: Clarendon, 1998.

Williamson, Tom. *Shaping Medieval Landscapes: Settlement, Society, Environment.* Macclesfield, Cheshire: Windgather Press, 2003.

Wolch, Jennifer, and Jody Emel, eds. *Animal Geographies: Place, Politics, and Identity in the Nature-Culture Borderlands.* London: Verso, 1998.

Wolfe, Cary. *Animal Rites: American Culture, the Discourse of Species, and Posthumanist Theory.* Chicago: University of Chicago Press, 2003.
———, ed. *Zoontologies: The Question of the Animal.* Minneapolis: University of Minnesota Press, 2003.
Wrightson, Keith. *English Society, 1580–1680.* New Brunswick, N.J.: Rutgers University Press, 1982.
Yamamoto, Dorothy. *The Boundaries of the Human in Medieval English Literature.* Oxford: Oxford University Press, 2000.
Zirkle, Conway. *The Beginnings of Plant Hybridization.* Philadelphia: University of Pennsylvania Press, 1935.

Contributors

Jeffrey Jerome Cohen is professor and chair of English at the George Washington University in Washington, D.C. He is the author of *Of Giants* (1999); *Medieval Identity Machines* (2003); and *Hybridity, Identity, and Monstrosity in Medieval Britain* (2006). He is the editor of *The Postcolonial Middle Ages* (2000).

Susan Crane is professor of English at Columbia University. She is the author of *Insular Romance: Politics, Faith, and Culture in Anglo-Norman and Middle English Literature* (1986); *Gender and Romance in Chaucer's* Canterbury Tales (1994); and *The Performance of Self: Ritual, Clothing, and Identity during the Hundred Years War* (2001).

Barbara A. Hanawalt is the King George III Professor of British History at Ohio State University. She is the author of *Crime and Conflict in English Communities* (1979), *The Ties That Bound: Peasant Families in Medieval England* (1983), *Growing Up in Medieval London: The Experience of Childhood in History* (1993), *Of Good and Ill Repute: Gender and Social Control in Medieval England* (1999), and *The Wealth of Wives: Women, Law, and Economy in Late Medieval England* (2007). In addition she has edited a number of books on medieval studies and has written articles on medieval social history.

Julie Berger Hochstrasser is associate professor of the history of early modern northern European art at the University of Iowa. While her research and publications to date have focused primarily upon seventeenth-century Dutch still-life painting, her current work investigates other forms of art and visual culture resulting from the global cultural interactions of the early modern period.

Richard C. Hoffmann is professor of history at York University in Toronto, Canada. Trained in medieval studies, he has evolved into an environmental historian of medieval and early modern Europe through his research, articles, and books on agrarian life, medieval frontiers, fish and fisheries, and urban ecology.

Joel Kaye is professor of history at Barnard College. His area of concentration is medieval intellectual history, including the history of science and the history of economic and political thought. He is the author of *Economy and Nature in the Fourteenth Century: Money, Market Exchange, and the Emergence of Scientific Thought* (1998). His current research centers on the emergence of a new model of equilibrium within scholastic thought, ca. 1225–1375.

Lisa J. Kiser is professor of English at Ohio State University. She is the author of *Telling Classical Tales: Chaucer and the Legend of Good Women* (1983), *Truth and Textuality in Chaucer's Poetry* (1991), and many articles on medieval environmental history, the history of animal/human relationships, and other aspects of the premodern natural world.

Pamela H. Smith is professor of history at Columbia University and the author of *The Business of Alchemy: Science and Culture in the Holy Roman Empire* (1994) and *The Body of the Artisan: Art and Experience in the Scientific Revolution* (2004). She has published numerous articles on artisanal knowledge and culture in early modern Europe, and in current research she is attempting to reconstruct the vernacular knowledge of early modern European metalworkers from a variety of disciplinary perspectives.

Marjorie Swann is associate professor of English at the University of Kansas. She is the author of *Curiosities and Texts: The Culture of Collecting in Early Modern England* (2001). She is currently at work on a book entitled *Without Conjunction: Desire, Society, and Anti-Fruition in Early Modern England*.

Index

Adam and Eve, 7, 42, 46, 50, 139, 145–47
Aesop, 43
agoutis, 193n33
agriculture, 2
Alan of Lille, 29n6, 42–43
Albala, Ken, 35n48
Albert of Saxony, 104, 110n30
Al-Biruni, 112n51
Albrecht V of Bavaria, 117
alchemy, 121, 128, 134n37
Alexander of Aphrodisias, 110n30
Al-Ghazali, 107
alluviation, 27
Alpers, Svetlana, 158
Anderson, Frank J., 9n4
animals
  as allegories, 45–48, 52, 55
  behavior of, 127–28
  depiction of, 155–99
  habitats of, 165
  and human boundaries, 38–57
  husbandry of, 5
  and hybridity, 52–57
  and identity, 39–62
  life-casts of, 119–20, 126–28
  naming of, 41, 58n4
  and race, 48–50, 54–55, 61n21
  reproduction of, 5, 140
  as symbols, 42–44, 45
Anna, Duchess of Bavaria, 121
anteaters, 193n33
ants, 165
Aristotle, 89–103, 107n7, 111n36, 151n13, 186
Ark, Noah's, 140
armadillos, 191n14, 193n32
Arnold, David, 9n5
Arnott, Michael, 62n31
artisans, 6–7, 116, 121, 128, 129–31
asses, 51–53
astrology, 101
Audoin-Rouzeau, Frédérique, 35n46, 35n49, 35n52
August I, Elector of Saxony, 116
Augustine, Bishop of Hippo, 74, 140
Avril, François, 81n16

badgers, 40, 41
Bailey, Mark, 32n27
Baillie, M. G. L., 32n22
Baker, Steve, 52, 57n1, 61n20
Bale, Anthony, 60n18
Balée, William, 30n11

228 Index

Balibar, Étienne, 60n21
Barlaeus, Caspar, 192n17, 195n53
Barlow, L. K., 31n20
Barnish, S. J. B., 32n23
Barry, R. G., 31n20
Bartholomeus Anglicus, 35n47
Bartlett, Robert, 58n4, 59n8
Bartosiewicz, László, 35n45
Baxter, Ron, 9n4
bears, 41
Beavan, Iain, 62n31
Bede (the Venerable), 48
bees, 45, 141, 143, 153n40
Beltran, E., 111n37
Bennett, Joan, 150n6
Bergvelt, Ellinoor, 131n2
bestiary, 5, 9n4, 45–48, 54–56, 60n15, 60n18
Biard, Joel, 108n9
Biddick, Kathleen, 36n58
Bildhauer, Bettina, 60n18
birds, 27, 41, 43, 44, 57, 121, 124, 165, 174, 182
Biringuccio, Vannoccio, 134n36
Blockmans, Wim, 132n7
boars, 6, 65, 69, 76, 78, 79
Boehrer, Bruce, 9n6
Boesch, Hans, 133n23–24
Boeseman, M., 190n7, 190n9, 192n24, 193n31, 199n83
Bordessoule, Nadine, 8on2
Bork, Hans-Rudolf, 37n71
Boswell, John, 60n16
Botkin, Daniel, 20n11
Bott, Gerhard, 133n23
Bourin-Derruau, Monique, 33n33, 37n70
Boxer, Charles Ralph, 190n6, 190n8, 190n10, 192n18, 194n36, 194n40
Brázdil, Rudolf, 30n14

Brazil, 7, 155–88, 189n3
Bredekamp, Horst, 116
Bresc, Henri, 33n33, 37n70
Bridgwater, Benjamin, 151n9
Brienen, Rebecca Parker, 189n3, 190n8, 191n15–16, 194n35, 196nn59–60, 197n62, 197nn66–67, 198n77
Brown, Neville, 32n2
Browne, (Sir) Thomas, 7, 137–50, 150n5, 151nn11–12, 152n18–19, 152n26, 153n30, 153n38, 153n40
Buridan, Jean, 6, 90–105, 106n1, 108nn15–16, 109nn25–26, 110nn32–33, 111n36, 111n42, 112n45
butterflies, 141
Butters, Suzanne B., 134n40
Buvelot, Quentin, 192n18, 196n57, 197n63
Bynum, Caroline Walker, 106

Caillois, Roger, 57n1
Camillo, Giulio, 118, 122, 128
Campbell, Bruce M. S., 37n68
cannibalism, 179, 182–84, 197nn72–73
Capon, Brian, 152n20
capybaras, 188, 193n33
Carmody, Francis, 60n14
carnivores, 20–28, 34n37
Carson, Rachel, 7
cashews, 170–73
Cassiodorus, 17
cereals, 25–26
Chaucer, Geoffrey, 40–41, 43, 78
Chimaera, 52
Chrétien de Troyes, 61n25
Chuine, Isabelle, 31n18
Clark, R. L., 37n69

Clark, Willene B., 9n4
Clavel, Benoît, 34n40, 35n53, 36n56
Clement of Alexandria, 46
Cleyer, Andreas, 162
climate, 5, 14–20
clocks, 92
coatis, 167, 194n41
Cohen, Jeffrey Jerome, 5, 60n19, 61n28
Cole, F. J., 152n19
collections of objects, 2, 6, 115–35
Collingwood, R. G., 4
Collins, Minta, 9n4
colonialism, 7, 168, 186
Comet, Georges, 36n60
cooking, 2
Cooter, William S., 37n67
Copernicus, Nicolaus, 102
Corrêa do Lago, Bia, 192n16
Courtenay, William, 108n16
cows, 41, 141
Craig, Michael, 57
Crane, Susan, 6, 57
Crawford, Patricia, 150n8
Creager, Angela N. H., 57n1
Crook, D. S., 37n69
Crosby, Alfred W., 4, 195n46
Crouch, David, 59n7
Crumley, Carole L., 9n5
Cummins, John, 35n55, 67, 80n5, 80n7, 82n18, 82n22, 82n27, 83n35
Curley, Michael J., 60n14

Dalché, Patrick, 99, 111n37, 112n47
Daston, Lorraine, 8n1
Davis, R. H. C., 82n20
deer, 6, 41, 45, 65, 72, 76–79, 79n1, 165, 169
Dekkers, Midas, 59n5
Delort, Robert, 4

dendrochronology, 18
Derrida, Jacques, 52–53, 55, 57
desertification, 19
Desportes, François, 185
Dickinson, William R., 30n13
diet, 20–28, 41
Digby, Sir Kenelm, 137, 142, 145, 149
disease, 20
Dobson, R. B., 62n32
Dodd, Anne, 34n41
Doering, Oscar, 134n27
dogs, 6, 58n4, 64, 67, 68, 71, 72–76, 79
Douglas, Mary, 78
drought, 18
Du Cange, Charles du Fresne, 106n2
Duhem, Pierre, 92, 93, 99, 109nn22–23, 110n28, 100n30, 112n47
Dunin-Wasowicz, Teresa, 37n72
Dunton, John, 151n9
Dutch West India Company, 155, 158, 168, 182, 192n18
Dyer, Christopher, 36n59

eagles, 124
Eckhout, Albert, 159–65 *passim*, 169–87, 191n16, 196n58, 196n61, 197n64
ecology, historical, 2, 4, 9n4, 13
Economou, George, 4
ecosystems, 25, 26
Eden, Garden of, 7
Edgmond, Florike, 191n13
Edward (of Norwich), Duke of York, 63, 65, 72, 76, 84n48
eels, 140, 141
Effmert, Viola, 133n26
Egbert, Virginia Wylie, 108n14
Eichberger, Dagmar, 132n7

Eiriksson, J., 31n20
elephants, 46
Emanuelsson, U., 37n66
Emel, Jody, 57n1
environmental history, 2, 3, 4, 5, 11–38
erosion, 94–96, 109n26
Ervynck, Anton, 21, 36n58
ethnicity, 5
Evelyn, John, 149

Farley, John, 153n28
Faust, D., 30n16
Ferdinand II, Archduke, 116, 117
Ferguson, Charles A., 83n39
Ferrão, Cristina, 189n3
Filotas, Bernadette, 35n50
Findlen, Paula, 116
Finke, Laurie A., 61n28
Fischer-Kowalski, Marina, 30n11
fish, 22, 24, 27, 140, 141, 165, 169
Fish, Stanley E., 153n40
flies, 140, 141
Flood, the, 140
floods, 18
Flores, Nona C., 43
folklore, 2
food chains, 5
forest law, 67
forests, 5, 9n5 24, 26
Fortune, 106n3
Fox, H. S. A., 37n67
foxes, 41, 71, 76, 167
Frederick III, King of Denmark, 185
Freedberg, David, 158
Frenzel, Burkhard, 31n17
friendship, 148–49
frogs, 127, 140
Fuciková, Eliska, 132n11
Fudge, Erica, 9n6

Gál, Erika, 38n73
Galen, 23, 89
game, 20–23, 25, 65–66
gardens, 195n52
Gasking, Elizabeth B., 151n10, 151n15, 153n37
Gassendi, Pierre, 153n28
Gauricus, Pomponius, 129
Geertz, Clifford, 64
Geoffrey of Monmouth, 48–49
geology, 93–105, 112n50
George, Wilma, 9n4
Gesner, Konrad, 4, 191n13
Gilbert of Poitiers (de la Porée), 29n6
Gildas, 48
glaciers, 18
Glacken, Clarence J., 4
Glaser, Rüdiger, 15, 32n27
Glass, Bentley, 152n19
Gleitman, Lila R., 83n39
goats, 22, 27, 34n39, 41, 144
Gobelin Tapestries, 185, 189n3
God, 7, 71–72, 91, 118, 124, 139, 143–46, 149
Grabmeyer, Johannes, 29n7
Grant, Edward, 4, 109n21, 110n27
grasslands, 27
Green, S. W., 37n66
Grew, Nehemiah, 141
Griebe, Jacob-Wilhelm, 185, 189n3
Grieco, Allen J., 34n27, 35n52, 36n59
Gries, Christian, 132n3
Grote, Andreas, 131n2
Grove, A. T., 32n21, 32n25, 33n29, 36n64
Grove, Jean M., 30n15, 33n32
Grove, Richard H., 58n2
Grzimek, Bernard, 193n28
Gudger, E. W., 190n8

guinea pigs, 169, 193n33, 197n63
Guiot, Joël, 31n17
Gulf of Lions, 27
Gunn, J.D., 32n24

Haberli, Wilfried, 32n28
Hanawalt, Barbara A., 57, 106
Haraway, Donna, 54, 55, 57n1
Hardouin de Fontaines-Guérin, 80n4
Hardy, Alan, 34n41
Harvey, William, 140, 151n15
Hassig, Debra, 9n4, 46, 59n10, 60n15, 60n18
Hastrup, Kirsten, 33n31
Hauschke, Sven, 133n25
Havenstein, Daniela, 146, 151n9, 151n11
hawks, 40, 144
Hayward, J.F., 132n10, 133n23
Helfferich, Emil, 199n84
Henry I, King, 41–42, 59n7
heraldry, 2
herbals, 4, 9n4
herbivores, 25, 26, 28
Herlihy, David, 29n6, 30n10
Highmore, Nathaniel, 140, 142, 151n17
Hill, Emily A., 84n41
Hitler, Adolf, 48
Hochstrasser, Julie Berger, 7, 197n67, 197n70
Hoffman, Richard C., 5, 35n51, 36n57
holly, 44
Holzhauser, Hanspeter, 32n28
Honour, Hugh, 197n73
Hooykaas, Reijer, 134n40
horses, 6, 65, 69, 71, 78, 141, 143
Howe, James, 71, 76, 82n22, 83n35
Howell, James, 149, 152n19
Huebert, Ronald, 150n7

hunting, 2, 5, 6, 23–24, 63–84
  contact with animals in, 76–79
  cries in, 65, 73–76, 82n33, 83nn34–37
  cutting up quarry in, 68, 84n49
  horn blasts in, 65, 72–73
  human/hound relationship in, 72–76
  as ritual, 63–84
  and the term "à force," 79n1
hunting parks, 24, 65
Huntley, Frank Livingstone, 150n5
Hüster-Plogmann, Heide, 34n43
hybridity, 5. *See also under* animals
hydrology, 27
hyenas, 46–47, 54–55

Impey, Oliver, 116

Jacopo da Strada, 116
Jacques de Brézé, 82n33, 84n49
Jacques Legrand, 111n37
Jamnitzer, Wenzel, 116, 119, 122–30, 133nn25–27
Jansen, Dirk Jacob, 128
Jews, 45–48, 54–55, 60n18
John of Salisbury, 49
Jones, Eric, 38n73
Jordan, William Chester, 57n1

Kauffmann, Thomas DaCosta, 116
Kaye, Joel, 6, 107n8, 108n12
Kemp, Martin, 116
Keynes, Geoffrey, 150n6, 152n22
Keys, David, 32n24
Kiser, Lisa J., 57, 106
Kistemaker, Renée, 131n2
Klingender, Francis, 4
Knapp, Ethan, 57
Knottnerus, Otto S., 33n33
Knudsen, K.I., 31n20

Krenkel, Paul, 29n7
Kris, Ernst, 134n41
Kruger, Steven, 60n18

Lac d'Annecy, 27
Lactantius (Lucius Caecilius Firmianus Lactantius), 151n16
landscapes, 161, 195n52
lapidaries, 4
Latour, Bruno, 8n1
law, 2
Laxton, 27
Leach, Edmund, 67
Le Blévec, Daniel, 33n33, 37n70
L'Ecluse, Charles de (Carolus Clusius), 159, 165, 191n13, 193n32
Leduff, Charlie, 60n19
Leveau, Philippe, 36n65
lice, 140, 143, 147
Lindberg, David, 4
Lingis, Alphonso, 53, 55
Linnaeus, Carolus (Carl von Linné), 159
Little Ice Age, 14–16, 18, 32n25
livestock, 18–19, 27–28
Locke, John, 185
Louis XIV, King of France, 185
lycanthropy, 49

MacAloon, John J., 64, 81n11
macaws, 174–76, 177, 178, 196n57
MacGregor, Arthur, 116
Makowiecki, Daniel, 35n44
malaria, 33n33
Mandeville, John, 157
maps, 118, 187
Marcgraf, George, 159–69 *passim*, 174, 177, 186, 187–88, 190n8, 191n14, 194n35

Marie de France, 42
marmosets, 193n33
marriage, 149
Marvin, Garry, 84n43
Mason, Peter, 199n84
Maurits, Johan, 157–58, 161–68 *passim*, 170, 176, 184–85, 186, 192n18, 194n35
Mauss, Marcel, 198n76
Maximilian II, Holy Roman Emperor, 122
McCulloch, Florence, 9n4
McGovern, Thomas H., 31n20, 33n30, 34n42
McNeill, John, 9n5
McNelis, James I., 79n1
Meadow, Mark A., 132n4
mechanical philosophy, 92, 99, 104
medicine, 2, 87, 119, 121, 190n10
menageries, 42, 59n7
Mentzel, Christian, 162, 163, 166, 192n21
Merchant, Carolyn, 4
Merton, Egon Stephen, 145, 152n26
mice, 40, 152n19
Mileham, Dorothy, 138–39
Mills, Robert, 60n18
mining, 12–13, 117, 123, 125
Mitchell, W. J. T., 5
monkeys, 39, 165, 193n33
Montanari, Massimo, 35n47, 35n49
Moody, Ernest A., 99, 109n17, 109n21
Moore, Sally F., 66, 71
Morineau, Michel, 32n27
moths, 141
Muñiz, Arturo Morales, 38n73
Muñiz, Dolores Carmen Morales, 38n73

Muratova, Xenia, 62n31
museums, 116
Myerhoff, Barbara G., 66, 71

natural philosophy, 4, 6, 85, 87, 90, 92, 95, 96, 104, 105, 106n1, 151n12
nature
   as category, 2, 3–4, 98, 101, 104
   criticism, 3
   equilibrium in, 6, 85–113
   generative powers of, 125
   human effects on, 1, 5, 6–7, 8
   imitation of, 126, 128, 130
   mastery of, 117–18, 119
   observation of, 155–99
Newport, Elissa L., 83n39
newts, 140
New World, 2–3, 7, 164
Niedenthal, Samuel, 185
Nile delta, 96
nominalism, 107n9
Nordenfalk, Carl, 84n45
Norman Conquest, 51
Norse settlements, 18–19

Ogilvie, A. E., 31n20, 33n30
Ogilvie, Brian W., 9n4
Oldfield, F., 37n69
opossums, 193n33
Orderic Vitalis, 42
Oresme, Nicole, 92–93, 104, 108n16, 110n30, 110n33
Ortolani, Franco, 30n16
Ostrów Lednicki, 22
otters, 41
Overton, Mark, 37n68
owls, 45
oxen, 140, 143
Oxford calculators, 91–92, 108n11

paca, 193n33
Pagliuca, Silvana, 30n16
Pálsson, Gísli, 33n31
panther, 45
Paracelsus (Theophrastus Bombastus von Hohenheim), 134n39, 141, 152n19
Parry, M. L., 33n32
Paster, Gail Kern, 57
Patrides, C. A., 150n3
Pearsall, Derek, 9n6
Pearson, Kathy L., 36n59
Pechstein, Klaus, 133n23, 134n28
Perlim, J., 195n46
petkeeping, 2
Pfister, Christian, 15, 31n17
Phébus, Gaston, 63, 65, 69, 72–78 *passim,*, 82n29
phoenix, 45
Physiologus, 45–46
pigs, 6, 22, 23, 24, 25, 35n52, 78. *See also* boars
Piskorski, J., 36n65
Piso, Willem, 158, 186, 190n10
plants, 141–45, 149–50, 152n21, 170–73
Pliny the Elder, 4, 122, 186
Pomian, Krzysztof, 131n2
population, 18, 24, 25
porcupines, 42, 194n33
Post, Frans, 159, 161, 181, 185, 187, 189n3, 192n17, 199n92
Postles, David, 37n68
Preston, Claire, 150n1, 152n17, 152n26, 153n40
Pretty, J., 37n68
Prevenier, Walter, 132n7
Procopius, 17
procreation, 137–54. *See also* animals, reproduction of

Pudsey, Cuthbert, 189n3
pumas, 194n34

Quiccheberg, Samuel, 115–31, 132n9

Rackham, Oliver, 9n5, 32n21, 32n25, 33n29, 36n55, 36n64
Ramus, Peter, 129, 130, 134n39
Rehazek, André, 34n43
Remigereau, François, 84n48
reptiles, 125, 127
Ribémont, Bernard, 103, 109n21, 112n47, 113n53
Richards, John F., 9n6
Riera-Melis, Antoni, 34n37, 36n59
Rivius, Gualterius, 134n41
Roberts, N., 30n13
Rogers, J. C., 31n20
Rogers, John, 151n11
Rooney, Anne, 80n7
Roosevelt, Theodore, 81n9
Ross, Alexander, 140
Rossi, Paolo, 134n40
rotation of earth, 102–3
Roth, Harriet, 131n1 *passim*
Rothfels, Nigel, 5
Royster, Francesca, 48, 60n19
Ruddiman, William F., 19–20, 34n35
Rudolph II, Holy Roman Emperor, 117, 122, 124, 133n23
Ruffus, Jordanus, 71
Ruhland, Florian, 36n58
Ruijsch, Frederik, 185
Russell, Emily W. B., 9n5
Ryff, Walther, 129

Saalfeld, Diedrich, 35n54, 36n62
saints, 65, 118
Salisbury, Joyce E., 4, 57n1, 67

Salter, David, 57n1
Salter, Elizabeth, 9n6
Saulnire, Chantal de, 80n2, 81n14
Scheicher, Elisabeth, 132n3
Schmalkalden, Caspar, 157–59, 163, 165–68, 172–75, 182–84, 187–88, 189n1, 191n12, 196n58
Schmidt, Benjamin, 198n82
Schmidtchen, Gabrielle, 37n71
Schnapper, Antoine, 131n2
Schneevogel, Paul (Paulus Niavis), 12, 13, 28
Scholasticism, 6, 85, 105
Schultz, Eva, 132n8
Schwarz-Zanetti, Werner, 31n17
Seba, Albert, 185
Sedgwick, Eve Kosofsky, 67, 71
Serreze, M. C., 31n20
sexuality, 5, 40, 41, 45–46, 137, 138–39, 151n11
Sharrock, Robert, 142
sheep, 22, 23, 27, 34n39, 41, 49, 61n24, 144
Shichtman, Martin B., 61n28
Shiel, R. S., 37n68
shrews, 40
Siddle, D. J., 37n69
Siewers, Alfred K., 58n2
Silvester, Bernard (Bernardus Silvestris), 11, 13, 28, 29n6
slave trade, 181–82
sloths, 7, 155–70, 173, 187–88, 193n28, 193n32
Smith, Jeffrey Chipps, 132n7, 134n39, 135n43, 135n45–46
Smith, Pamela H., 6
snakes, 127. *See also* vipers
Soares, Jose Paolo Monteira, 189n3
soils, 14–15, 26–27
Sonnlechner, Christoph, 36n60

Index 235

Sorabji, Richard, 5
Sörlin, Sverker, 10n7
Spiegelman, Art, 61n20
sports, 2
Staden, Hans, 197n73
Stanford, Michael, 151n11
Steel, Karl, 57, 59n9
Steinberg, Theodore, 30n12
Sterry, Peter, 153n34
stoats, 40
Stock, Brian, 28n1
Strathern, Marilyn, 3, 8n1
Strubel, Armand, 80n2, 81n14
Swann, Marjorie, 7
Switsur, R., 30n15
Sylla, Edith, 108n11

tapirs, 163
Tapuya Indians, 176–83, 186, 196n61, 197n64, 197n71, 197n73, 199n84
Taunay, A., 190n8
Teixeira, Dante Martins, 168, 169, 186, 189n3, 191n14, 193n31, 194n41, 195n42, 195n45, 196n61, 198nn78–79, 199n89
theology, 2
Thiébaux, Marcelle, 80n5, 80n7, 82n18
Thiessen, Erik D., 84n41
Thomas of Chobham, 42
Thomas, Keith, 4, 195n46
Thomas, Marcel, 81n16
Thorndike, Lynn, 111n37
Tilander, Gunnar, 82n31, 82n33, 83n34, 83n36
transformation, 51–52. *See also* hybridity; lycanthropy; witches
trees, 18, 26, 126, 137–54. *See also* forests
trophic pyramid, 20–21

Tupinamba Indians, 176–77, 182, 185, 199n84
Turner, F., 195n46
Twiti, William, 63, 65, 73, 76, 83n37

unicorns, 46

van Berkel, Klaas, 8n2
Van den Boogaart, Ernst, 189n3, 190n8, 194n35, 196n59, 197n64
van Helmont, Johannes Baptista, 152n19
van Kessel, Jan, 185, 188, 189n3, 191n14
Vanderjagt, Arjo, 8n2
Vartanian, Aram, 151n16
Verhulst, Adriaan, 36n60
Vidal, Fernando, 8n1
vipers, 45, 46
Vitruvius (Marcus Vitruvius Pollio), 122
Vives, Juan, 134n39
Voisenet, Jacques, 9n4
volcanic eruptions, 17, 32n25
von Schlosser, Julius, 131n2

Waddington, Raymond B., 150n2
Wagener, Zacharias, 168–70, 174, 179–82, 189n3
Wallerstein, Immanuel, 60n21
Walter, Francois, 9n5
war, 2
Warde, Paul, 10n7
Watanabe-O'Kelly, Helen, 132n3, 133n12
weasels, 40, 46
Weiss, Gail, 56
Welsh, bestial nature of, 48–49
Westra, Haijo Jan, 29n1
Wetherbee, Winthrop, 28n1
Wheel of Fortune, 106n3

Whitehead, P.J.P., 162, 190n7, 190n9, 192n24, 193n31, 199n83
Wilding, Michael, 139
Wilhelm V, Duke, 117
William of Conches, 29n6
William of Malmesbury, 41–42, 50–53, 61n28
William of Moerbeke, 107n7
William of Ockham, 107n9
Williamson, Tom, 9n6, 35n51, 36n61
witches, 51–52
Wolch, Jennifer, 57n1

Wolfe, Cary, 5, 48, 57n1
wolves, 24, 167
Worm, Olaus, 185
worms, 140
Wrightson, Keith, 153n41

Yamamoto, Dorothy, 9n6, 67
Yapp, Brunsdon, 9n4
Yiou, Pascal, 31n18

Zielhofer, C., 30n16
Zirkle, Conway, 152n21
zoos, 186

www.ingramcontent.com/pod-product-compliance
Ingram Content Group UK Ltd.
Pitfield, Milton Keynes, MK11 3LW, UK
UKHW020402240525
458856UK00006B/146